In Service of the Wild

In Service of the Wild

Restoring and Reinhabiting Damaged Land

Stephanie Mills

Beacon Press
Boston

Beacon Press
25 Beacon Street
Boston, Massachusetts 02108-2892

Beacon Press books
are published under the auspices of
the Unitarian Universalist Association of Congregations.

99 98 97 96 95   8 7 6 5 4 3 2 1

Text design by Janis Owens
Composition by Wilsted & Taylor

Library of Congress Cataloging-in-Publication Data

Mills, Stephanie.
In service of the wild: restoring and reinhabiting damaged land /
Stephanie Mills.
    p.   cm.
Includes bibliographical references and index.
ISBN 0-8070-8534-0
1. Restoration ecology—Case studies. I. Title.
QH541.15.R45M55   1995
639.9—dc20   94-38094
                                                       CIP

To

Sally Van Vleck and
Bob Russell

and to
Richard Nilsen

I take infinite pains to know the phenomena of the spring, for instance, thinking that I have here the entire poem, and then, to my chagrin, I hear that it is but an imperfect copy that I possess and have read, that my ancestors have torn out many of the first leaves and grandest passages, and mutilated it in many places. I should not like to think that some demigod had come before me and picked out some of the best of the stars. I wish to know an entire heaven and an entire earth.

H. D. Thoreau, *Journal*

Contents

Acknowledgments

Thanks to Barbara Dean, whose luminous friendship and literary wisdom have sustained me through many an endeavor; thanks also to Barry Lopez, whose encouragement to attend a Society for Ecological Restoration conference kindled my interest in writing on this subject and whose fraternal support in a couple of doubt-racked moments kept me writing. Thanks as well to the E. F. Schumacher Society, whose invitation to give one of their annual lectures caused me to concentrate my attention on restoration, and to produce the nucleus of the present work.

Without help from individuals and institutions, this work, and the journeys that constituted it, would not have been possible. I am deeply grateful for the generosity of Hildegarde and Hunter Hannum, Ramona Morgan, and my parents, Robert and Edith Mills. A grant from the Foundation for Deep Ecology, and the auspices of the Neahtawanta Research and Education Center were fundamental to this project; to both of these visionary organizations I am indebted.

Steve Packard very kindly reviewed the Schumacher Lecture and gave graciously of his time and encouragement on several occasions as the project progressed. Dave Egan commented thoughtfully and helpfully on the book proposal. Kelly Kindscher tutored me on prairies, and exemplified how wide and rich botanical wisdom can be.

Max Holden and Bill Herd at the Sleeping Bear Dunes National Lakeshore helped launch my research into the pre-Columbian history of my life-place. Roger Williams shared his understanding of the Grand Traverse region's land-use history, and Bill Brunelle enlightened me on the subject of property lines and title searches. Susan Stachnik shared her story of life on this land. John Petoskey and Lewis Sawaquat spoke with me about the history of native people here; James McClurken and Charles Cleland shared their knowledge and provided scholarly materials on pre- and post-settlement Indian history.

The following individuals were good enough to review the manuscript whole, or in part. Any errors remaining are my own responsibility, while the merit of the work rests greatly on the intelligence of: Tency Baetens, Nina Leopold Bradley, Matthew Bremer, Joss Brooks, Rob Chapman, Bob Doherty, Dave Egan, Tor Faegre, Don Falk, Allen Feldman, Susan Flader, Chip Francke, Fred Fuller, Karen Furnweger, Ed Giordano, Alan Herbert, Phil Holliday, Freeman House, Bill Jordan, Virginia Kline, Roger Klocek, Kraig Klungness, Rick Moore, Janet Morrison, Steve Packard, Brian Price, Milo Pyne, Laura Quackenbush, Bill Rastetter, Chuck Robbins, Laurel Ross, Guy Ryckaert, David Simpson, Tineke Smits, Ted Sperry, Randy Stemler, Bill Sullivan, and Seth Zuckerman.

Deanne Urmy, my editor, and a darn fine bookwoman, deserves much credit for making this work work, as does Sue Flerlage, who is no mere typist, but a lodestar of sanity.

Finally, it is impossible adequately to thank, however I may try, Charlotte Robertson, an incomparable friend whose careful review of and improving suggestions for the second draft of the manuscript were timely as the gift of a fairy godmother; and Tom West, whose dear presence in my life has restored it to sweet, sweet love.

Prologue

Is it perhaps peculiar to cherish difference any more? To resist simplification and homogenization, whether of the landscape or of the human psyche, goes against the grain of industrial civilization. Yet people do have an appetite for variety and rarity in Nature. There probably wouldn't be so many maniacal bird-counters if that were not so. There's just something about a species; a unique response to circumstance; a syllable of beauty. What our materialism has done is to rip the figure from its ground, the creature from its habitat. Hence we don't quite grasp the whole implication of rarity. A rare bird has to call some place home; as does a rare blossom or a great big mammal. Peculiar beings can't exist without peculiar surrounds; they give rise to one another.

All species that are now scarce and endangered once were plain members and citizens of their ecosystems; they may once have been plentiful. The great ecologist Paul Ehrlich, always handy with a quip, likens the random and ever-increasing elimination of species on our planet to a random removal of rivets from a plane in which we're flying—it's not possible to predict precisely which extinction will precipitate a major disaster, but every one increases its likelihood.[1] Hence the essentiality to all beings, ourselves included, of biodiversity, and of the habitats that sustain it. Protection is step one—eliminating the immediate threats to the sacred place, or ecosystem. Rallying people to the mighty struggle for wild- and open lands preservation remains critical, and the nineties will be the make-it-or-break-it decade for that effort. Protecting unspoiled lands will be no mean feat. Beyond that, taking some meaningful action against the planetary threats—such as depletion of the stratospheric ozone layer, and contamination of the biosphere by man-made chemicals with their insidious effects on whole suites of organisms—is crucial also. But none of those reactions suffices as a basis for culture.

The archbioregionalist Peter Berg once likened the environmental movement to a hospital that was all trauma center and no birthing room, then went on to say that bioregionalism sought to be that birthing room; to bring forth something hopeful in the midst of all the injury. Protecting the lands that have been saved, enlarging preserves through restoration, and restoring the everyday landscapes in between, will be the agenda for the third millennium. Restoration is about accepting the brokenness of things, and investigating the emergent property of healing. It's the closing of the frontier—ceasing our demand for open land to "develop"—and the reinhabiting of exploited or abandoned places. Facing these necessities and doing the work liberates human energy; yields a bounty of knowledge and satisfaction, and a resurgence of wild Nature. Restoration is a wide opportunity for enjoyment in the land, a sense of serving the sacredness of Nature, and touching it with your hands.

In Service of the Wild is about restoring the Earth, ecosystem by ecosystem, watershed by watershed. It is about regaining a sense of how beautiful, how stable, how hospitable to an abundance of life forms and human cultures healthy ecosystems can be. It is about individuals and groups that have done, or are doing, remarkable service in their ecosystems. It's about an epochal change in our relationship with the land, our species beginning finally to take an interest in attending to what the land itself has wanted to bring forth, its creation of self-regulating communities of organisms, climax ecosystems. "It's not about control, but about surrender," say the prairie restorationists in Chicago. "The biosphere is using this river to pour all this information into our hearts," say the salmon restorationists in Northern California.

Ecological restoration is the art and science of repairing damaged ecosystems to the greatest possible degree of historic authenticity. The discipline's fidelity to the original ecological communities of the places being restored is a profound obeisance to Nature. There is a tremendous range of restoration activity—many hundreds of projects across the United States and abroad in a wide variety of ecosystems, reefs, salt marshes, arid farmlands, prairie potholes, Alpine meadows, mangrove swamps—the list is about as long as the list of the kinds of ecosystems that have been wounded by human activity.

Restoration involves all manner of work with the land, from removing roads and restoring the contours of terrain to removal of exotic plant and animal species that are eliminating or out-competing native species; to planting trees, grasses, and wildflowers; to propagating endangered plant and animal species. Much of this work, although conceived by professionals—ecologists and land-use managers—is performed (and informed) by volunteers. Restoration skills can be developed by anyone, and wielding them can produce psychic benefits at least as great as their beneficial effects on the land.

In the land we may find solace for our wounds, privacy for a developing intimacy with a natural surround, an occasion for acting out healing processes that effect inner healing as well; or we may remain unconscious of and oblivious to the living community of the land. Numbed and paralyzed by the degree of damage that has been inflicted on the land, we may be domineering and exploitive toward it, or even blindly destructive. Our behavior toward the land is an eloquent and detailed expression of our character, and the land is not incapable of reflecting these statements back. We are perfectly bespoken by our surroundings.

The transformations of ecosystems by settlement have been noth-

3

ing if not radical. And yet, until quite recently, historians all but ig-
nored the changing natural history within which human stories have
been played out. Environmental history, which addresses the ecological context for human events, has wide relevance here. It is a means
of reconstructing an image of earlier landscapes—the image to restore
to—and it is a detailing of the countless acts, inventions, and imper-
atives that, whatever the intention, tore apart the land with breathtak-
ing speed. Moreover, it is a study of cultural beliefs about land. In
environmental history we may find some of the clues we need to begin
to remedy our collective psychological alienation from the Earth that
bears us.

I am a landowner, although not of the self-reliant homesteader
variety. Even so, my possession of my land, and my developing intimacy with it, coincides with and has animated the making of this
book. I am, I guess, a kind of fundamentalist. I was suckled in the
sixties' creed. Going back to the land was about far more than dropping out. It was, and is, about finding a big enough relationship in
which to be fully human. I have begun and ended this book on my
acres in Leelanau County, Michigan. This is where I *have* to learn
about the land. I take myself as a specimen of *Homo sapiens reinha-
bitens*, a person in need of all manner of knowledge of and functional
concern for her home ground. So I have devoted the first part of the
book to chapters on my quest to see places in my bioregion that still
retain fragments of their original ecosystems and that therefore exhibit
the quality of land health. I've gone on to the story of the deconstruction of the ecosystem I might have inhabited had I lived here before
Columbus, and finally to a kind of diary about learning to see and love
my land as it is now.

My purpose in telling the history of my home ground is to evoke
for the reader a sense of how dramatic and profound the changes
wrought in even pleasant rural landscapes are; how much biodiversity
and land health has been lost, and why the *ecological* dimension of
restoration is so necessary. I also would convey how inadvertent the
degradation was: nobody consciously sets out to wreck a piece of land,
but that is a common result of our accustomed habits of land use.

The second part of the book tells five restoration stories. It moves from the domain of the individual landholder (an extraordinarily influential one—Aldo Leopold), to an academic institution, to an urban volunteer program, to a rural reinhabitation effort in the Pacific Northwest, to an international utopian community doing landscape-scale restoration in the Third World. Although not quite a neat set of nested wholes, each succeeding example has broader and more radical implications for the future of humanity in Nature.

Aldo Leopold was a landowner who through his writing addressed his whole culture. A pioneer in ecological restoration, as well as in wildlands preservation, Leopold, too, owned some derelict farmland in the upper Midwest. As a householder he set about to heal it. Some of what he did is described in *A Sand County Almanac*. Perhaps more important, this great book argues the *why* of restoration and preservation, for changing "the role of *Homo sapiens* from conqueror of the land-community to plain member and citizen of it."[2] Leopold's Sand County farm, now part of the Leopold Memorial Reserve, certainly qualifies as the birthplace of a philosophy of restoration in America, and was irresistible to visit. So Aldo Leopold's Sand County farm, as it is today, is where the second part of the book begins, with a look at what the man and his family were able to accomplish in the way of land healing, and at the work that continues there today.

One of the wonderful things about being a nonfiction writer (and you wouldn't think there were any, to hear all the griping that goes on around here some days) is that it provides a brief to satisfy your curiosity by asking questions of informed sources. You can talk to learned professors and academics without having to suffer through all the tedium of a graduate education. An interview is a tutorial, and that's what I got during my visit to the University of Wisconsin at Madison's Arboretum, which has been, since Aldo Leopold's time at the university, a center of research in the techniques of ecological restoration. It also serves as the home base for the Society for Ecological Restoration, the interdisciplinary group that is leading the movement, promoting professional standards of quality in the work, and encouraging a deeply philosophical consideration of the work's meaning.

One of the techniques developed at the Arboretum—prairie restoration—has been famously implemented in an urban setting. Chicago, Illinois, is where prairie restoration has become a civic pastime;
where the practitioners and organizers have developed a rhetoric that
is luscious and persuasive. It is a setting where thousands of people
work as volunteers to restore their native prairie and oak savanna landscapes, often within city limits. These projects are conducted with the
guidance of conservation organizations and the Cook County government, but they are essentially the work of the volunteers who tend
nature preserves scattered across the metropolitan region—or "Chicago Wilderness," as some of the organizers would have it. It's a vivid,
brilliant manifestation of the "Green City" or "Wild in the City"
ideas advocated by urban bioregionalists, a beginning of the reinstatement of Nature's authentic presence in large human settlements.[3]

From the Midwest, I return to Northern California, where I lived
for fifteen years of my earlier life and work in the ecology movement.
As an ecology activist in the seventies, and an heir to the sixties' sensibility of radical dissatisfaction with attempts at reform, I was easily
persuaded that environmentalism, reformist at base, could at best only
be a holding action. Mainstream environmentalism didn't much dare
to question our power arrangements, either intra- or interspecies, and
only sought to sanitize civilization. It was in the San Francisco Bay
Area where, owing to a 1970s encounter with then-Diggers Peter Berg
and Freeman House, I would learn about bioregionalism. (The San
Francisco Diggers were a band of anarchists serving the half-transient
community in the Haight-Ashbury with free food, free clothing, and
street theater. The name *Digger* paid homage to a small band of
seventeenth-century English communalists who, during the era of en-
closure, attempted to cultivate, or dig, the wastelands. They planted
a commons in Surrey. Their community was short-lived and was destroyed by mob violence fed by the antagonism of adjacent landowners.)

The bioregional vision emerged clearly in the mid-seventies,
made a deeper analysis of ecological crises and proposed more satis-
fying, *devolutionary* solutions. Ongoing friendship and colloquy

with Peter Berg, Freeman House, and, following the first North American Bioregional Congress in 1984,[4] a host of other, equally creative and ecologically literate *beaux esprits*, led me to become an exponent of bioregionalism and an aspiring reinhabitor.

Peter Berg is one of the geniuses of bioregionalism, and, with Raymond Dasmann, the author of the idea of reinhabitation. Berg and Dasmann explained reinhabitation as follows:

> If the life-destructive path of technological society is to be diverted into life-sustaining directions, the land must be reinhabited. *Reinhabitation* means learning to live-in-place in an area that has been disrupted and injured through past exploitation. It involves becoming aware of the particular ecological relationships that operate within and around it. It means understanding activities and evolving social behavior that will enrich the life of that place, restore its life-supporting systems, and establish an ecologically and socially sustainable pattern of existence within it. Simply stated it involves becoming fully alive in and with a place. It involves applying for membership in a biotic community and ceasing to be its exploiter.[5]

For the last twenty years Freeman House has been trying to imagine what reinhabitation might be. For nearly that long, with his colleagues in the Mattole Watershed Salmon Support Group, House has been laboring in some very hard circumstances to save a species, and engaging in some rigorous thinking and utterly lucid speaking and writing to explain why. A visit to the remote Mattole River valley in Northern California, including a long conversation with House, and conversations with other leaders in Mattole salmon restoration, was a journey on par with the pilgrimage to Aldo Leopold's farm.

The philosophical shift from restoration as an avocation and abiding passion in a metropolis, to an attempt by a group of "New Settlers" to restore a watershed's capacity for natural provision is the subtext of the Mattole story. For more than a decade the salmon restorationists have been exerting themselves in cold rivers and around heated community meetings to create conditions favorable to the endangered races of salmon native to the Mattole. In that their work is meticulous, and pivots on the fate of an indigenous creature, it qualifies as ecological restoration, certainly. But it was undertaken with a view toward restoring the circumstances necessary for a human population to subsist, contentedly and indefinitely, in this particular place. And those conditions necessarily include the development of a local consensus about the way of life in the Mattole watershed, and eventually, a good many residents hope, the evolution of a culture that fits this place and its diverse and sometimes contentious groups of inhabitants.

If we accept that reinhabitation is the proper and necessary vocation for all twenty-first-century people, then it must get under way in multicultural communities (which most human settlements of any size now are) and it must go on in the Third World as much as in the First and Second Worlds. (The Fourth World, indigenous peoples, continue to inhabit their life-places; no need to begin as postmoderns. Their struggle is to maintain their land bases and their subsistence lifeways.) Accordingly, the final example in the book is Auroville, a visionary intentional community in southern India where followers of the philosophy of Sri Aurobindo are regenerating 2,500 acres of blasted earth to sustain what they hope eventually will be a fifty-thousand-strong "city that the Earth needs." Aurovilians began twenty-five years ago by busting through the hardpan with crowbars to plant trees, and thus far have got about half their land in healing mode. They also have been working in appropriate technology development, and are embarked on realizing a Utopia that includes biology.

The fact that Auroville is an intentional community is salient. Just as ecological restoration strives to reestablish native plant and animal communities, reinhabitation hopes to reestablish the ageless human capacity for forming self-reliant communities. It hopes to rekindle our evolution-bestowed genius for what the gentle anarchist Kropotkin called Mutual Aid. The forces that are tearing apart ecological communities are identical with those tearing apart human communities, and there is probably no saving the former without regenerating, or sometimes reinventing, the latter. Industrial civilization has claimed far too much diversity—both biological and cultural—already. We live in the moment of NAFTA and GATT: insane, Leviathan proposals, totally antithetical to the preservation of humane and ecological values. Resisting their effects will be an epochal struggle.

There are a million ways to do it, and they all have grass roots. This is not to say that global understanding isn't necessary. But the countercultures resilient enough to endure the upheavals caused by the imploding too-much and the heartlessness of transnational capitalism, the countercultures and indigenous cultures resilient enough to survive the occasional collapse of a degraded ecosystem, will have to be modest in scale and adamantly devoted to restoring their life-places to health.

The idea of restoration raises the problem of encouraging humans to imagine that any damage to any landscape can be healed, and of creating an ecosystem-managerial elite and industry. Restoration, narrowly construed, could just be another form of greenwash. It could be co-opted, limited to natural landscaping on corporate and superpower campuses or freeway offramps. There is also the possibility that it will fail, just as other, earlier, ruder attempts to enhance ecosystems—like German forestry—failed over time. Or that global- and atmospheric-scale problems will obviate restoration attempts. If restoration is not done, however, it seems certain that many more species of plants and animals will be lost; also that the possibility of re-establishing big wilderness, complete with big wild animals, will be lost forever. Finally, if we make no commitment to restore and reinhabit our life-places, our species, having come to a turning point in its moral evolution, will have taken a last wrong turn—away from community with all beings and toward a managed, synthetic environment. We can begin again to work with Nature, or continue to work against it (but if the latter not for very much longer).

In this troublous time, ecological restoration work represents a gallant gesture of sodality with our fellow beings here on Earth, an effort to counter the trend of extinctions, and the sincerest possible expression of concern for life's future. It has the eros of connection, of the physical. Ecological restoration, allied with the creation of ecosystem-scale wilderness reserves, represents the main hope that the organic quality of wildness may someday be resurrected in human souls and in all life-places on planet Earth.

It's not that human works and human society as we presently know them are not interesting; they just aren't interesting enough. There must be something else, something larger, wilder, more mysterious than just our species free to display its creativity and astonishments on our planet. Never mind the survival arguments, the likelihood that the viability of Earth's ecosystems could be all but destroyed (which means us with it, of course), forget about that apocalyptic alarum. Forget the bang and forget the whimper. Forget those times spent as a child huddled under your desk at school in training to survive a nuclear attack. Set aside, for a moment, the possibility that our current era of enterprise and conflict could precipitate a desolation that would qualify as an Ending Time.

Remember, instead, how good it is to wander in the woods, or the meadow, or down by the river; how it feels to be baking in the desert or sinking in the swamp or creeping through the forest or standing with the wind in your face in the big middle of the plains. Remember how it feels to be alert to movements in the variegated shade of the savanna. Remember what a relief it is, what a joy, to find yourself sentient in the company of wild things, secure in the knowledge that the beauty that surrounds you also has extent enough for its health, and your own. And if you can't remember, imagine it. Imagine the night being not silent nor humming with traffic, but rich with sounds of birds and frogs and insects, with stirrings of nocturnal mammals, all manner of vitality. Human sounds, yes—sighs, whimpers, breathing, whispers. Just an absence of engines and metal upon metal, wheels upon pavement. No crowds but crowds of blossoms. Imagine a world where the life of the Earth and of the human spirit could go on, evolving, diversifying, adapting, changing, and surprising, fearlessly: if it can be imagined, it can come to be. If it can be recalled, it may be restored.

Part One

Chapter 1

The Wild & The Tame

Late in August 1993, in the throes of writing, I had the strangest dream: I had to cut in horizontal thirds a copy of my first book in print, a memoir whose title asked, but did not answer, the question "Whatever happened to ecology?"[1] I applied a bread knife to the task. When I opened the three pieces, I discovered that the bottom third consisted of the blank space below the type. Then this fraction became a whole book, looking very much like Peterson and McKenny's *Field Guide to Wildflowers*.

You can't not think about a dream like that. What could it mean? The following evening I wandered out back in the field with my scissors and my cat, wanting a free bouquet. But the *gaillardia pulchella* I had planted on the slope were all done with their flaming ember show. Horsemint, a goldenrod, black-eyed Susans, and pearly everlasting were the only flowers remaining, and none of them appealed to the florist in me. Spotted knapweed being abundant, but *planta non grata*, got a pass, too. I did, however, find a feather striped in russet and walnut, tipped dark, then white, that looked as if it might recently have belonged to a hawk; and saw that the likely fox den had some fresh excavation. Sign of fox, sign of hawk, traces of the wild.

In the field a possible interpretation for the dream occurred to me. To explain it requires me to take two leaps backward in time: first, to 1970, when I had titled the never-to-be-finished tome I was working on "Whatever Happened to Ecology?" I was at that time a fairly well known young ecology activist, and I thought the environmental movement was forsaking Mystery for policy, wildness for respectability—selling out, in short. Hence the confrontational tone of the question.

In 1985, when, having retreated to the country to attempt the simple life, I took up work on a to-be-finished book of the same title, I had a less pat answer, and wound up writing a personal story, something of an apology for pat answers. *Whatever Happened to Ecology?* was a narrative of my journey into the bioregional movement. Bioregionalism is decentralist, place-located, ecosystem-based. Part of the answer to that open question, then, was that ecology and social change had become, for bioregionalists, mutually informing if not inseparable.

Leap forward to September 11, 1992. It was my birthday, the day of a full moon. I was in "Little Tibet"—Ladakh, now the northernmost part of Jammu & Kashmir State in India. A conference called "Rethinking Progress" took me there, to Leh, the capital city.

Given some free time, a fellow conferee and I took off for a village outside of Leh to consult Amalamu, a prospering oracle and healer. The séance was held in her kitchen. There was an apprentice with her, and although the way they went about inducing their trances was expert, the shuddering, yipping, and convulsing cannot have been easy work.

My question to Amalamu was as convoluted as only an introvert's could be, having to do with whether the nature of the service I am called to do is artistic or political. It proved to be untranslatable in detail and got boiled down to: what should I do? What *do* you do, she wanted to know. I write about Nature, I said. Keep it up, she said.

"If you write a book about past, present, future of Nature, your future will be bright."

Naturally it was nice to hear about the possibility of a bright future—standard soothsayer talk, I thought. But I saw no obvious reason for her to advise me to think about the *future* of Nature. Perhaps two-thirds of whatever happened to ecology—its past and present—are already set in type, and they cannot be revised or rewritten. But the last third, the future of Nature, is marginal, open, a blank page at first, then an entire book, a guide to the wildflowers, a field guide. Perhaps even a "feeled" guide.

Like it or not, we are entering a new era in Earth's history where wild Nature may persist at the macroscopic level only by human sufferance.

In a valedictory address to the Society for Conservation Biology, one of the Society's co-founders, Michael Soulé, described the prospect with some irony:

> Global warming, drying, ozone depletion, and toxification will produce many unpredictable ecosystem-level effects that will have dramatic impacts on biodiversity. Fragmentation will continue to harry habitats; its bitter fruits will provide many opportunities to discover and treat new kinds and degrees of area and edge effects and related maladies. The truncation of biological communities by the removal of top carnivores, important herbivores, plants that provide critical resources, and habitat-reforming taxa (beavers, elephants, pocket gophers, termites, etc.) will yield valuable information about the roles of keystone species. Restoration ecology and conservation biology will tend to merge because most so-called wild places on the planet will be relatively denatured and will need intensive rehabilitation and management.[2]

I came into the ecological concern at a time when the valuing of wild over tame was the rock upon which conservationists built their church. Thoreau's saying "In wildness is the preservation of the world" was a basic article of faith. I became an armchair devotee of that holy wildness, understanding it to be the ground of all Earthly being, including the human. Yet one of the most startling realizations of the present moment is that the living tissue of wildness—biodiversity—is utterly dependent on our care. The "our" of our care as of this writing includes 5.4 billion human beings, a number that is growing. Wilderness buffs *may* number in the thousands; ecologically concerned individuals, ecosystem peoples, and members of traditional subsistence cultures perhaps in the millions. This is but a small portion of the population alert to the fundamental issue. Thus there is a whole lot of information exchange and consensus building—call it reinhabitory work—before us if we are to secure the future of Nature. Every kind of effort has got to go on simultaneously, from solitary artisanal work creating prairies to promoting what the educator David Orr calls "ecological literacy,"[3] to hard-core struggles to protect what remains of old-growth anything, be it forests or coral reefs.

The psychological and moral shock of assuming our existential responsibility for the preservation of species is terrific. To me it seems as though the ground of mind, god, and being is shifting. For humans to locate themselves in right humble relation to wild Nature under these changed circumstances, at the end of an era whose metaphors derived from mechanics and which promoted the illusion of control, will require a radical rededication to the wild. The future of Nature will be a function (albeit not an exclusive one) of the soundness of human community. We might prefer it otherwise but must concede that our species holds the hole card now. Conservation biologist Reed Noss has observed that our species is more adaptable (and more destructive) than any other.[4] Our mixed blessing of hyper-adaptability must now be put into service, to restore the whole of Nature, the health of Nature, which is to say, wildness. And our species—so terribly alienated from the wild, so displaced from native community—is in as urgent need of healing as is the land.

In the summer of 1993 unusually heavy rains deluged the upper Mississippi River Basin, and a so-called natural disaster—devastating inundation of the agricultural lands claimed from, and of the cities built on, the river's floodplain—wrought havoc in thousands of human lives. The economic effects of this disruption of our concentrated large-scale production of agricultural commodities, and of providing at least some government relief for the individuals and communities whose lives were nearly destroyed, will likely be rippling through the nation for years to come.

Much of the public reacted by blaming the river, damning Nature, and spoiling for another fight with the vast region's hydrology. There was some mention of the inevitability of flooding in floodplains. There was little mention of how the deforestation of the watershed's uplands, the draining of its lowlands, the displacement of deep-rooted prairie ecosystems by countless acres of pavement and almost-as-impermeable lawn, the compaction of the soil by farm equipment, all had, in truth, helped to create the disaster. These floods were partly the result of what is deemed necessary human activity, activity undertaken with a degree of ecological obliviousness, if not plain ignorance. There were, as well, people throughout the region who, well aware of the water's occasional propensity to rise, loved living by whichever tributary so much that they planned to stay put, or to return as soon as their trailers drained out, with cussed affection for their lately wild watercourse.

A few voices were suggesting that perhaps, if by now the Army Corps of Engineers had not succeeded in getting the river to behave rationally, it might be time to negotiate, and to allow some wetlands and floodplains to revert from human purposes to performing their timeless function of buffering the periodic changes in streamflows. Thinking like this might even portend a boom in restoration activity (as might the breakdown of so many other outsized attempts to redirect natural systems for exclusively human purposes, be they military, economic, or agricultural).

Even when attempts at domesticating, or subjugating, wild systems or organisms appear to have succeeded, the results are a little perverse and pettish. (Consider the passive gluttony of hatchery salmon, the flavorlessness of store-bought, midwinter tomatoes, and the foot-wetting neuroses of miniature poodles.) In contrast to these anomalies are the qualities of wildlife and wild places: authenticity, indigeneity,

17

fierceness, and spontaneity; resilience and health above all. The ecologist-philosopher Aldo Leopold wrote of *land health* ("that capacity for internal self-renewal")—a quality he deemed rare even in his youth, almost a century ago. Land health is what we behold in wilderness (its "perfect norm," said Leopold), perhaps what we go to wilderness to behold.

There are two threads here: wilderness and reinhabitation. Wilderness, Americans who enjoy such surroundings have come to think, is rocks and ice; high, remote, challenging places. This is because at the beginning of this century such austere, albeit gorgeous, leavings of land—then apparently unprofitable and, by humans, uninhabitable remnants of the public domain—were what could be acquired for parks and wildlife refuges. But wilderness—wildness—is less a matter of vacancy and topography than of biological diversity and abundance. In wilderness, which until Columbus was what the New World was, all the land—prairies, floodplains, deserts, and forests—could be fat with life, with lives, both in numbers and kinds, and even could accommodate a sparse scattering of aboriginal peoples.

"The wilderness," Nancy Newhall wrote, "holds answers to more questions than we yet know how to ask."[5] In the train of "natural disasters" (which often are the failure of human enterprises to be resilient within the planet's incessant changing), those questions become questions of survival, and not merely survival of our own species. Culture historian Thomas Berry avers that if human consciousness had evolved on the moon, it would be as barren as the moon.[6] As the ongoing industrial crusade to turn all Earthly life to commercial purpose relentlessly impoverishes the biosphere and human cultures, our living images of graceful possibility dwindle.

What little wilderness remains displays the patterns we must restore to, if our species and as many others as now remain are to persist here a while. Ideally this would call for a broad cultural *rapprochement* with the wild, a long-overdue armistice in civilization's war upon it. Reinhabitation means learning the whole history of one's bioregion or watershed, and developing a vision of sustainable ecological community from that knowledge, and from what we have been learning, in the last half-century, about elegant techniques of construction, gardening, recycling, energy conservation, and waste treatment; ways of sophisticating old-style household and neighborhood frugality.

In addition to ending the hostilities to wilderness by reining in our material wants and needs, there must be a surrender of conquered territory. Conservation biologists now propose that for there to be sufficient habitat to allow the survival of North America's ecosystems and the species they comprise, at least half of this land would have to be removed from human use and restored to a wilderness condition. Substantial wilderness reserves are the heart and soul of the bioregion, epitomizing its character, existing as its sacred ground. Concentric zones of increasing human use and settlement density would embrace the wilderness reserve.[7] Livelihood would be patterned on, and derived from, the stability and economy of the climax ecosystem. This posits a postindustrial human culture and a postmodern indigeneity.

I have no doubt that the valiant engineers and flooded-out farmers of the upper Mississippi River Basin knew some of the history of their region, possibly even of other, earlier agricultural regions around the world that were finally bested by their rivers and rebuked by their exhausted soils. Probably the engineers thought they had no alternative but to try again, and imagined, perhaps, that our remarkable postwar technologies would enable them to get it right this time. Today the possibilities in cooperating with—rather than conquering—the great natural systems seem much richer and more widely practicable. Restoring and reinhabiting bioregions will not be the province of a managerial elite, but of what we all do indefinitely, mindful that the subsistence lifeways of land-based tribal peoples worked well for millennia and left only modest traces on the land.

The hope of ecological restoration has begun to produce a subtle Earth change in my sense of the future. It all depends on our consciousness, on an informed and deliberate willingness to proceed into a millennium in which our long-accustomed relationships to Nature are reshaped, and in which we ourselves are reshaped by our dealings with Nature. One still hopes for a liberatory future, a far gentler, indeed a happier future. I believe that restoration will help us there. My faith is tied, I admit, to a mystical faith in the epiphanic value of ecosystems. Thoreau's insight—that in wildness is the preservation of the world—takes on new significance in our time. Now it expresses why we must revert to Nature. We cannot learn freedom and responsibility within the confines of our own species. We cannot understand life and death and what they are for in exclusively human terms. Without that which is wild, the world becomes a cell block.

There is a quality of wildness in every living organism, in every great planetary process—weather, tectonics, cell division, instinct. These are phenomena of ineffable grace. A few years ago, I stood on a beach with watershed-organizer Freeman House, watching the Mattole River meeting the Pacific Ocean, wondering aloud what was to become of wilderness. I had always somehow envisioned the ocean in such an encounter as passive, just receiving the river's water. Here the sea broke back on the river mouth; there was an embroiling of the waves and the river, of salt water and fresh. When the steelhead trout are running, there are seals and sea lions just off the bar, hunting in the surf, and clouds of seabirds hovering over them. "The air is wild," said Freeman. "Your intestines are wild." Life and its emergent properties are wild. The soul is wild. Certain kinds of human gatherings— affinity groups, watershed councils, intentional communities, committees working by consensus—give forth ideas and relationships that are novel, transcendent, surprisingly greater—more compassionate and livelier—than the sum of each person's imaginings, and hence demonstrate a kind of wildness. Some essence of wildness will persist in us unto our last breath. The microbes that return us to our elements are wild, too.

The fact of this irreducible essence of wildness does not qualify the duty of ecological restoration, which, like conservation, is to ensure that in the future wildness will be multifarious and on all scales, including the continental; that there will be a wild variety of habitats and species and that the fullness of evolution may proceed. The struggle to allow this is now in its decisive throes. In addition to admiring those who know the plants and grow the trees, I have utmost respect for wilderness defenders and traditional indigenous peoples striving to preserve their homelands. They are a dedicated and necessarily intransigent kind. If it were not for their determination to fight for the sanctity of unspoiled places, and to keep on fighting despite having pieces of their hearts torn out with each loss, we would have today no hope of knowing wildness writ large and whole. There would be no refuges for beings larger than ourselves, no inedible or natural landscapes, no expanses great enough to shelter the shy and finicky creatures of the deep interior. Human concerns, important as they are, pale into repetitive melodrama alongside this epochal necessity to relinquish our pursuit of dominion over all of the Earth. If as a species we can consciously evolve enough to preserve and restore wilderness, then wildness will secure for us a world.

Civilized humans, caught in the solipsism of human supremacy which is reinforced everywhere by our increasingly artificial environment, have a great deal of difficulty grasping the magnitude of what is going on at this climactic moment of life's history on Earth. We're all in this web together, and our very human nature depends on that quality of richness and complexity in the living fabric that enfolds and sustains us. Accordingly, habitat restoration and the reintroduction of extirpated species become important human occupations, tributary to this larger process of giving Nature room to move. "Evolutionary diplomacy," Freeman House calls it.

Development of techniques by which to do this—to safeguard, for now, the genetic endowment of native species; to rescue and propagate them wherever they are—is a transformational practice. It sits within a larger cultural and political task, to which Aldo Leopold alluded in remarks made to the Wildlife Society in 1940: "To change ideas about what land is for," he said, "is to change ideas about what anything is for." It means changing our ideas about what *we* are for: is ours to be a dominant, parasitic relationship with all other species, or coordinate? It's the big ontological question and we can't balk it.

To be honorable, restoration activity must be inseparable from the prevention of further habitat destruction. Otherwise it misrepre-

21

sents the founding fact of specificity, which is that places are different. The plant associations that are the basis of life and livelihood for most animals, ourselves included, arise where they belong and belong where they are. Microclimate, microhabitat, differences in soil composition, all give rise to genetic variation even within species. Thus an ecosystem is a firmly rooted thing. The possibility of certain kinds of ecological restoration—establishing new patches of prairie plants, for instance—should not lure us into the delusion that ecosystems can be moved around like oriental rugs. Nor should we make little plans: the top carnivores and omnivores which are the truest sign of an ecosystem's health and wholeness are cosmopolitan and wide-ranging. Their main habitat requirements may be room to move and freedom from human harassment. What must be restored to ensure a future for the wolf, the grizzly, the cougar, and other great predators is *extent*.

If we fail in preservation and restoration, we default into an ersatz world where people have to get their kicks in virtual reality because Earthly reality has become desolate. In assenting, however unconsciously, to the destruction of biological diversity, we domesticate ourselves and constrict our experience. Wild and free are much the same, and are the antithesis of domestic. Wilderness preservation and ecological restoration are, on different scales, homage to undomesticated creatures and communities. They hinge on authenticity and indigeneity. Prairie restorationists in Chicago try to limit their gathering and propagating of plant matter to populations that can be found within fifteen- or twenty-mile radii of the sites being restored. Advocates of wolf reintroduction in the Sonoran bioregion of southeastern Arizona and northern Mexico consult federal records on wolf sightings so that their proposals for migration corridors will be authenticated by natural history. Salmon restorationists in Northern California work for a race of fish native to their river, not for hybrids, clones, or next-best.

This loyalty to the minute particulars of DNA in native species stands in contrast to the genome craze of our time. Conservationists and restorationists are interested in genes as they are spontaneously and integrally expressed in organisms, for the plain fact of their being.

22

Genetic engineering continues the mania for predictability, control, and exploitation; the conquest of the tiniest redoubt of the wild. Genes are wilderness, too; submicrocosmic wilderness, made of, by, and for whole places.

One winter night I had a conversation with a friend in which we were both attempting to draw the boundary between natural and what's come to be called *virtual* reality. Making this distinction challenges even the bigtime pro philosophers, thus it is not surprising that we amateurs wandered so bootless.

My friend was trying out the proposition that everything that is, is in a sense a product of evolution; that by that token virtual reality is reality and that bionic limbs and computer-chip replacements for areas of the brain, by dint of functioning as part of an organism, will come to be regarded as living.

I ransacked my logic for arguments against this sacrilegious proposition; such artifacts are not alive because they lack the capacity for self-healing, because their manufacture entails pollution, because they have no emergent properties, because, in the case of computer chips in the brain, they will bring forth no surprises. Nothing in a category that has not already been thought of will be conceived by them.

As the conversation forged on, I was privately, uneasily aware that my revulsion toward the realm of robotics and cybernetics, toward virtual reality and artificial life, has left me willfully ignorant of some doings that may indeed blur the distinction between natural and artificial. There really are roboteers who think it would be an advance to replace our brain tissue with computer circuits. There are researchers working on "manufacturing" processes that could take place at the atomic level—nanotechnologies—that may amount to being a kind of engineered metabolism. The manipulation of the genetics of organisms in order to maximize one or two of their qualities—like milk production in cows or frost resistance in strawberries or antibody production in cells—is going forward at a brisk clip. There are engineers and scientists devising machines that can propagate and repair themselves, and computer software that "designs" organic, unpredictable forms—artificial life, they call it. That the term *life* is now applied to things apart from biology is a measure of how rapidly this audacity has proceeded.

Such hubris spawns bad accidents, accidents that spill out and engulf the innocents—creatures quietly minding their own business, living out their evolutionary destinies, each having an integral relation to the whole. I worry about any optimism that tempts us to become careless about all these other actors of fate. Casually we propose to go forward, to go Nature one better: to design plants immune to pests, animals that produce like factories, machines that mimic cells, and people invulnerable to death. The human mind can be so lethally self-interested: "autistic" is how Thomas Berry has characterized the last few centuries of our relationship to the Earth.[8] It shakes out as a religious question, I guess—and the question is, "Is nothing sacred?" Are there no natural phenomena—cells, organisms, ecosystems—before which we might just stand in awe, forfending them from "improvement"?

As our conversational play with abominations and integrities wore on, I became short-tempered with my friend. Finally, I withdrew, frustrated. I had not been able to forge a purely intellectual argument for the sacredness of Nature and against the profanity of industrial civilization. Nature *per se*—ancient forests, prairies, termite mounds, and hollow wing bones—ought to be argument enough for humility. Why this is not self-evident eludes me (hallmark of a true zealot), but I guess it has something to do with the shape of my yearning for the holy. Much of the pleading of ecological activists in the twentieth century has been to let trees fall in the forest—it matters not whether soundlessly—and rot: for the sake of the soil, for the sake of *that* sacred community. It is a pleading that the value of Nature is intrinsic and need not, indeed should not, be counted only in terms of what humans derive from it.

This is a hope, not a theory: if increasing numbers of us participate in ecological restoration where we live, we will learn with our flesh the difference between natural and artificial. Through positive engagement in ecosystems, we will come to revere evolution's fine and canny design processes so greatly that we will stop at nothing to preserve all remaining species and the habitats they need.

What restoration could and should be for *in us* is the transformation of our souls. In addition to what this work may accomplish in the land, I yearn for it as the yoga that will cause us to evolve spiritually, that will restore to us a feeling of awe in something besides our own conceits. I fervently hope that this work—in our streets, yards, fields, woodlots, parks, creeks, and watersheds—does indeed hold the potential to carry us into a postmodern, postindustrial, postcivilizational relationship with Nature. I pray that it is our way back into the web of life, carrying with us what we know now.

Talking with Roger, a visitor from Wisconsin, one July morning, I told him that on his weekend tour of Leelanau County he should bear in mind that every place he'd be going—fields and farms and open country—there once would have been an old-growth hardwood forest, with huge trees, even a few six-foot-diameter elms scattered through. In relating this land history to Roger, who is, among other things, a farmer, I gained an appreciation of the pioneers' investment of toil in clearing this country. I got a sense of the muscle in the heroic saga of opening ground for agriculture. It is not so difficult to understand why growing food should be felt to be unequivocally good, and why the price paid for that in human sweat should be honored. Agriculture and its resultant civilization have for millennia now been upheld as representing a greater good than savagery in the wasteland. So these patterns not only dominate our economics and culture, they constitute much of the structure of myth.

I've always chafed at the idea that a new myth is required for us to make it into the next millennium, without razing the biosphere. The deliberate creation of an authentic myth seems so paradoxical— but perhaps myth is one of the dimensions in which restoration must work. The Gnostic author Stephan Hoeller proposes that "the restoration of biological order must be seen as a metaphor for the restoration of the wholeness of consciousness; otherwise we shall miss the deeper meaning of the present crisis of civilization.'" Healing is the dominant metaphor in ecological restoration (as in so many realms of the current human drama). Restoration's leading exponents and theorists speak in those terms. Some of them even hint that the healing process is as much shamanistic as scientific—that a ritual sense should inform the work.

Restoration, and reinhabitation, might become a way of dealing with our all-too-human problem of impotence and dominance in the face of Nature. Dominance has been encoded in our thinking about everything, from how we treat one another to how we treat the living world. Cooperation, attentiveness, and partnership (even with subtle, unseen forces)—the values that prevail in restoration and shamanistic healing and effective community—are perhaps more difficult to mythologize because they cannot be drawn in terms of individual heroics, victorious battles, and subjugation of the wilderness. Indeed it is especially such myths from that Western-ethnocentric complex that must give way to something more conducive to restoration. Our tutelary images could, once again, come from Nature—plant and animal powers guiding and allying with, sometimes correcting, human conduct.

Most people on Earth are emigrants and immigrants, more or less recently. Lately the pace of deracination and the quest to establish new home places has intensified. Perhaps this quantitative difference in pace equals a qualitative difference. It spells an end to historic indigeneity, and a need for new eras of becoming native-again-to-place, of reinhabitation. Restoration, I think, needs to be considered one of the ways and means of reinhabitation, along with ecological agriculture, renewable energy development, and self-governance—all crucial to the future of Nature.

As I worked on this book, something that I originally envisioned as a minor element—environmental history, Nature's past—showed itself to be at least as relevant to the subject of healing as stories of ecological restoration. Retrospect, the study of my own life-place's history, caught me up for the better part of a year. The past of a place and of its people speak of its character and suggest the direction of a future autochthony. Indeed, that late twentieth-century society might course-correct and rebuild its communities in tune with the rhythms and customs of definite places—watersheds, bioregions—seems less farfetched to me than the idea that another bureaucratic technofix might be made to work.

During this epoch of sedentarization and civilization, our cultures have become increasingly effete, dependent on artifacts—both political and technological—that afford rude survival for the many and luxury for the privileged few. I think we will be weaning ourselves from this infrastructure for the next few centuries as we explore what it might be to be "future primitive."[10] The fact that the human population is now so immense, and that so much of the planet's habitat has been stripped of biodiversity, does not mean that we can't or shouldn't imagine a future for ourselves in our ecosystems, maybe even a nomadic future, a wild, postcivilized future. It's a few generations further down the road than ours. We won't live to see it fully realized, but even just beginning it might feel better.

My ultimate hoped-for outcome from restoration is not unremitting opportunity for high-minded horticultural toil, but an eventual return to an ecosystem balance (which implies a much smaller human population) and with that, hunter-gatherer proportions of leisure and dream time. Time for the kind of social play that results in a diversity of cultures, cultures as varied as the ecosystems that sustain them. Accordingly, one rubric for right action might be to ask whether my work today is contributing to the restoration of natural and cultural diversity, toward natural subsistence for future primitives. Is what I am doing contributing to the knowledge that individuals require to live independently and harmoniously and harmlessly on Earth, or is it leading to more sameness, more dominance, more dependency, less joy, fewer species?

Community can grow, embedded in restoration endeavor, galvanized by wilderness preservation struggles, fostered by concerted human action in service of the health of the land. Barn raisings made people feel good a century ago. Restorationists' salmon release and prairie burning provide new satisfaction today. Who knows what restorations of joy there might be a century hence? Buffalo runs down

Vincennes Avenue in Chicago? Topsoil and tigers and native rice again in Tamil Nadu? Grizzlies ranging from Mexico to Newfoundland? Waterfowl rising from the upper Mississippi tributaries and prairie potholes to darken the skies and weigh down a few tables? If we are writing the future of Nature, as Amalamu counseled, it better be bright, and no less prodigal than Earth's own creative genius.

Chapter 2

Scraps of the Virgin Cloak

In attempting to know what Thoreau termed "an entire earth," I've sought that knowledge both in prospect—by looking at the practice and possibilities of ecological restoration, which may allow us to amend much of our damage to the Earth—and in retrospect. The link between these two parts of this work is as much temporal as it is logical. During the time I was roaming afar to learn the ecology of restoration, I was also venturing into books and distant hardwood forests to get eyes to see back through time, to envision what my land, or land like it, might have been when wild. At the same time, I was venturing about on my acres to learn their modern, pidgin ecology, and bearing it all in mind, pondering a best purpose for my tenure in this particular place.

This chapter, and the two that follow it, recount that process of exploration and imagination, of coming to know something about habitat. The learning and telling takes time, yet making this prologue—a specimen history of place—seems essential if the hope is not merely to restore, but to reinhabit the land. Reinhabitation requires a sense of the origins of the soil, and of the human history that was played out upon it, and changed it in the process. Learning that one's natural surroundings may be pretty but lacking in ecological wholeness, and how that paradox came about, is part of building the kind of companionable relationship to place that can sustain a future of healing those surroundings.

This place where I live got its start five billion years ago with the coalescence of the hot, meteor-struck, cloud-swaddled Earth. Then there was rock and sea; two billion years ago something algalike lived in the vicinity of Michigan's Upper Peninsula (and elsewhere, no doubt), due to be fossilized, exhumed, and dated an unimaginably long time later. There was a proliferation of single-celled and multi-celled organisms. By 600 million years ago, Lower Michigan and much of the Great Lakes region was a basin in the stable craton that was the nucleus of North America during the Paleozoic era. For much of the Paleozoic, Lower Michigan was undersea. The seas receded and advanced; the sun still rose and set every day. The Earth's orbit and axis meant seasons of some sort in these north latitudes, and the northern lights must have shimmered even over lands devoid of animals and flowering plants, over coral-reefed seas plied by primordial fishes. The gray constellated Petoskey stones that Michigan beachcombers prize are corals 350 to 400 million years old—fossilized *Hexagonaria percarinata* from the middle Devonian period, occupants of one of the great salt bodies of water that from time to time covered Lower Michigan. Racing through evolution at a whirlwind pace, skipping lightly past the Mississippian and Pennsylvanian periods, which in Michigan were marine millennia, but on unsubmerged land saw the genesis of primitive plants and the ancestors of the conifers, shooting through the Mesozoic era, rampant with dinosaurs, onward into the Cenozoic, our age of mammals, we arrive, breathlessly, in the Pleistocene epoch.

There came a series of glaciations, eras of ice miles thick. The concavity that became Lake Michigan had its shape changed repeatedly. Lake levels rose and fell, shorelines shifted, vegetation advanced and retreated as the glacially slow pulse of the ice sheets decreed. This country was scraped bare by ice, denuded of life, planed raw and remade.[1] "If you would see a beautiful peninsula, look around you," advises the Michigan State motto. My home is on the Leelanau Peninsula, the "little finger" of the Lower Peninsula of Michigan (which is shaped, roughly, like the back of your left hand). The Lower Peninsula is flanked by two of the Great Lakes—Lake Michigan, 307 miles long and 923 feet deep at its deepest; and Lake Huron, 206 miles long and 750 feet deep at its deepest.[2] The numbers barely begin to give a sense of the immensity of these inland seas which, by moderating extremes of temperature, and providing abundant lake effect snows, condition so much of the life of the land.

It's pretty new around here. The fourth, most recent interglacial period only began some ten thousand years ago. As the ice retreated, first tundra, then grass, next spruce and fir, then pines, and finally the broad-leaved trees of the hardwood forests, succeeded onto the newly thawed land. Over a few thousand years the Leelanau Peninsula began to acquire its Great Lakes Forest community.

Nowadays as I travel my vicinity, I try to read the terrain as describing glacial action. This is drumlin and moraine country, drumlins being hills of glacial till, shaped and oriented by the flowing ice, recessional moraines being sinuous drifts of various-sized cobbles, gravels, and sands deposited by the edge of a melting glacier. Moraines mark pauses in glacial movement. Two miles to the east of my place are gravel pits—signs of glacial outwash—deposits of rock stratified and size-sorted by water melted and flowing from the ice. On my land, the soil is sandy, the crop of stones slight. A few great boulders also were transported by the ice, for the glaciers were eclectic in picking up and dropping rocks of different weights and gauges. These big stones are winningly called "glacial erratics." Often they are beautiful, mysterious, angular granites veined with dikes of quartz, and a good many of them are given places of honor in front yards around the county, an autochthonous touch.

In Michigan I have become acutely aware of living in an interglacial period, of inhabiting a dynamic landscape that was, until relatively recently, under ice and then was stark tundra for quite a few centuries after that. There were villages at Jericho before the Great

Lakes Forest grew up.[3] It wasn't until about 3000 B.C. that the flora and fauna of this region had come into their present association. (This we know thanks to the science of palynologists, who examine and differentiate pollen grains preserved in bogs or sediments. The varying frequency of different kinds of pollens in different strata discloses the changing membership of the plant community over millennia.)

No sooner had the land a sufficiency of time to recover from the big geomorphic disturbance effected by the glaciers than it was occupied by people, who became the next great agents of change in the biota. Paleo Indians were in this region long before the hardwood forests, establishing villages along Lake Michigan's shores. (And in this broad region, Paleo Indians [ca. 11,000–8000 B.C.] were probably a factor in the extinction of mastodons and woolly mammoths.) Following the spruce-fir forest, the emergence of the deciduous forest with its greater complement of species provided a richer, more dependable subsistence base, hence larger, denser human populations could develop at the forest's edge.[4] The Archaic peoples of this region (who were here from 8000 to 1000 B.C.) found rich subsistence in the elk, deer, bear, and smaller mammals; in fowl, shellfish, and turtles; and in the nuts, seeds, and other produce of the woods and waters.[5]

The deliberate introduction of foreign plants, notably corn, which originated in central Mexico, was well under way by the middle Woodland period (around 300 B.C.–500 A.D.). Beans and squash were also cultivated and provided a measure of stability in a hitherto variable food supply.[6] The first American missionaries to encounter Odawa and Ojibway people on the Leelanau Peninsula noted that they grew potatoes (which plants had made a long journey from their native Peru, possibly arriving via Jesuits from France, trading with Hurons). One could reasonably speculate that there may have been some movement of medicinal plants throughout the Great Lakes region, as various peoples of the northeast woodlands were displaced westward following the Iroquois ascendance. It may be that the abundance of certain plants was influenced as they were harvested, or not, according to the customs of the different cultural groups inhabiting these woodlands over time.

The Odawa people who inhabited this area at the time of European settlement hunted in the wintertime for meat and furs, gathered maple sap in the spring, and spent their summers near the shores of the lakes farming, foraging, and fishing, especially for whitefish and sturgeon. Although there's no way of really knowing what the interior experience of a hunting and gathering person, especially of the female persuasion, was in my bioregion, I imagine there was more to it than just awaiting the dropping of the fruits of nature into one's outstretched hand. More likely those hands were busy in summertime weeding rows of corn, beans, and squash, plucking berries, digging medicinal roots, gathering berries, cleaning and smoking fish, mending clothing and nets, weaving rush mats, and making birchbark containers. In winter there would have been cleaning game and working with hides, tapping sugar maple trees, some rest and singing, and no small anxiety about making it through to spring. There was no exemption from the need for skillful means, and no small amount of tending the "virgin" forest and waters, whether that tending meant setting fires to favor the growth of berries or tending of a spiritual sort, as in the ritual propitiation of the *manitos* of the lakes and streams.[7]

Some people talk about the theoretical possibility of recreating an extinct species from the DNA on up. Deplorable as I find that technofantasy, I must admit that my own approach to developing an image of the climax ecosystem of the place where I live—its tree and wildflower species, its birds and mammals—has been no less an intellectual thing, not laboratory- but library-based. Early on, I thought there could be a typical portrayal of these woodlands. As real information began to endanger that Platonic ideal, I realized that somewhere in the back of my mind had been a natural history museum diorama of a standard and predictable community of plants.

In *Deciduous Forests of Eastern North America*, E. Lucy Braun writes of the "physiognomy" of the various forest types. Her exhaustive scholarly devotion to the deciduous forest must surely have been motivated by a love of these faces. Her rigor and thoroughness and consistency make her book a sturdy tool. Yet the study is like trying to understand the flickering expression of a lover's face encumbered by the arcane preconceptions on a physiognomist's chart, complete with predictions of character. Really to understand the countless phenomena of the forest—the character of each bird, the meaning of its song, the medicine of every plant, the signatures and habits of predators and prey, their nesting, birthing, and mating—would take all the senses of everyone in the village for many generations. Part of what emerges from more recent works is that plant communities and forest types, the archetypal climax condition, are dynamic associations rather than fixed inventories. Add to this the baleful fact that global warming threatens to move vegetation ranges faster than vegetation can move, and the paradigm to which to restore blurs.

"Eastern Hardwood Forest" is a general tendency varying according to the particulars of place. The Great Lakes Forest is one such regional variant. Composition varies with latitude and altitude, slope, soil, and drainage. An Eastern Hardwood Forest in south-central Ohio will have a slightly different aspect—hickories along with beeches and maples, an earlier spring and longer summer, a later, shorter winter—than an Eastern Hardwood Forest in Michigan's Upper Peninsula, which is underlain by some very old rock and, for much of the year, overlain by some very deep snow.

The Great Lakes Forest won't stand still for a snapshot. Different creatures inhabit it at different times of the year, birds come and go; different wildflowers emerge by turns as the canopy above them becomes leafier and more dense and eventually shades them out. Differences in the ages of trees and in the tree populations of the forest are brought about by disturbances ranging in magnitude from the death of a single tree to the storm-caused windthrow of hundreds of acres. In areas where edges have been opened, which by now is almost everywhere, the composition of the hardwood forest has been affected by deer browsing, which hampers reproduction of the trees whose seedlings the deer are partial to, such as white pine, Canada yew, white cedar, hemlock, and yellow birch.

Sugar maples, the dominants in this forest, can be astoundingly prolific and patient. Millions of ankle-high seedlings may carpet a few acres, abiding the shade for decades, growing almost imperceptibly, awaiting some windfall of light. Of those millions of sugar maples, only a few will attain their full maturity—which might mean a four-foot girth and a seventy-foot bole rising straight up before branching out at eighty or ninety feet. In a time-lapse forest movie spanning centuries or millennia, you would see the constant winnowing of lives—the thousands upon thousands of maple and beech seedlings in just an acre being decimated repeatedly until only a few score remained. Extinction is forever but this culling kind of death means futurity. Spring after spring, blooms of delicate color would glimmer across the forest floor, as would the brighter buttery yellows of bellwort and trout lilies and the gaudy spangles of the large-flowered trillium.[8] Along with the green blades of wild leeks you might see hepaticas, bloodroot, violets of several colors, lady's slippers, Solomon's seal, jack-in-the-pulpit, baneberry, trailing arbutus, wintergreen, twinflower, crane's bill, all blossoming and fading, their bulbs or seeds blanketed through the winter by fallen leaves and sifting snows. An old tree falls, its roots wrench out of the dirt. In its descent its trunk may take down some neighbors. More light for the contenders, and a lens of raw earth opened, inviting a little garden. Millions of such events, a slow chaos, guarantee variation on a fine scale.

Incorrigible optimists like to mention that there's more woodland in New England today than there was 100 years ago, owing to the succession of old fields back into forests, and perhaps to wrest from this the hope that the Eastern Hardwood Forest is poised to regenerate itself everywhere. However, one must also allow that without elk, moose, caribou, wolf, cougar, and lynx, or without the *Homo sapiens* part of the ecosystem—indigenous subsistence cultures—those woods are a little thin. The primeval forest of eastern North America was undoubtedly spooky, just the thing to alarm a puritan imagination, and cry out to be leveled. Some of it was described before it fell. Here in the "Old Northwest," the fur-trading French voyageurs found little to remark about the virgin forest, perhaps because they had no intention of staying. They traveled mostly by water, and the

Indian settlements they visited were on watercourses and confluences, or near the lakeshore, so the vast timberland of Michigan's Lower Peninsula lived in obscurity until the movement of settlers up from the south began.

On a map of the Leelanau Peninsula, with its vegetation types noted according to the remarks of the first western lands survey circa 1839, the maple-beech-hemlock association prevails almost everywhere except in the wetter places, where cedar and tamarack predominate. Closer to the sandy lakeshore there would be pine and oak forest with blueberries and huckleberries in the understory.[9]

Thousands of years' residue of what Rutherford Platt the elder calls "the deciduous idea"—endless tons of fallen leaves, which unlike conifer needles are very quick to decay and turn their nitrogen back into the soil—would have, abetted by the fine questing threads of the fungae, provided a spongy, living tissue of the forest floor.[10] (In the summer of 1992, the gee-whiz science story in the headlines was the discovery of a giant "mushroom" in the Upper Peninsula, a vast [37 acres], ancient [1,500 years], heavy [100 tons] body of *Armillaria bulbosa* fungus that was proved by genetic analysis to be an individual, one enormous forest organism, and a fabulous statement of Nature's tendency not to be discreet.)

On these sandy, morainal hills, the subterranean welter of tree roots and root hairs would stabilize the soil and provide "routes" for water to move down into the ground and then back up into the atmosphere via evapotranspiration out the leaf surfaces. To think that these 300-some square miles of the Leelanau Peninsula were completely cloaked in trees! Unbroken forest cover must have meant a different hydrologic regime (the water table dropped following the logging, and many flowing streams vanished, hindsight testimony to the function of vegetation in the dynamics of a watershed), and a different sort of climate: moister, cooler, stiller, evener.

In virgin deciduous forest there would be mossy seeps, springs, and bogs, little kettlehole lakes winking back at the sky from their break in the summer canopy, or in winter forming an irregular plaza of flat white gripped by ice while all around the bare branches of the trees would veil the skyline in gray lace. The patterns of bark on tree trunks are clues to the trees' identity; there is the coarse pewter furrow of sugar maples, the evenly ridged hide of red oaks, the cat-scratch parallels of Eastern hop hornbeam, the neat gray corduroy of white ash, the dingy curls of yellow birchbark, the inky little shakes of black cherry, the cement smoothness of the beech, the white birch, lurid and papery—unmistakable.

A springtime walk in these woods has a buff undertone. Green ferns do not overwhelm you, but you might expect to find them here and there. The mat of leaves underfoot is tan and yielding, on drier days crisp and noisy. In the spring rains leaves underfoot are slippery, the duff under them is sticky. Forest clings to the feet, the feet inevitably crush a few flowers. Different plants with different appetites for sunlight (these ordained by the competition and cooperation that shapes the structure of the forest plant community) flowering at different moments of the season, of the day—a floral clock. Sometimes it is cold white and dead still, at other times muggy, green and rowdy with bird song, still other times crisp, blazing and rustling with scarlet and gold, gleaming with berries, rich with beechnuts. There is a time when the spent sugar maple flowers rain down and a time when poplar fluff—wind-dispersed seeds—begins to waft around on the summer breeze. There's a time when maple samaras, winged elm and linden seeds flutter down from the canopy as they ripen in their turn over the course of the year.

The uncut forest must have been a paradise for its four-footed inhabitants. There would have been elk, and moose, and even some caribou. Porcupines and their only menace, the fur-bearing fisher, would have contended in the depths of these woods, flying squirrels would have glided in the upper stories. Black bears would have been in this forest, wolves, cougars, lynx, and snowshoe hare. Great flocks of turkeys and passenger pigeons would have foraged on the beech mast and all the berries. Martens and wolverine preyed here. Habitat destruction claimed the big hoofed mammals, the fur trade would have taken out the fur-bearers, and implacable settler hostility claimed the remaining big predators.

Compared to the trees, animals are short-lived as mayflies. By a combination of great genes and good fortune, a bear or elk might live twenty or thirty years, whereas a sugar maple or beech may live for three or four centuries. Trees are extremely slow and sedentary. They don't flee at human approach, much as they might wish to, so mature trees are the most conspicuous members of the forest, also the most enduring. They're among the last organisms to perish in the ecosystem's decline. In the living memory of old-timers around here there were woods that were just plain too deep and scary to enter. Now it is necessary to go some distance to see a virgin hardwood forest of any size, and these persist mostly by dint of preservation efforts.

In my quest to see old-growth hardwood forests, I went with my young friend Rachael to the Colonial Point Preserve in Cheboygan County, about forty miles to the northwest. Saved for all time by the Little Traverse Conservancy with the help of the Nature Conservancy and the Michigan Land Trust Fund, Colonial Point is a forest of red oak, white pine, and other hardwoods grown to full maturity. Vanishingly scarce, the clear, straight dense timber produced by old forest trees becomes ever more valuable: the oaks of Colonial Point barely escaped a trip to Germany to become veneer.

The presence of a stand of red oak, which wants more light than the sugar maple and beech that dominate the region's forests, suggests that there was human management of these woods for centuries. The oaks in Colonial Point are all of an age—a century plus. The fact that they belong to roughly the same generation is a clue that they established themselves during an interval of anthropogenic fires during the mid- to late 1800s. Evidently the Odawa who occupied Colonial Point (until the 1900 eviction-by-arson of their village) farmed there, and maintained their fields by burning.

Indians of the eastern woodlands generally engaged in a fair amount of land clearing, primarily to open the ground for agriculture. Without metal tools, Indian tree removal was laborious and probably resulted in some fairly ugly vistas. Fires were set at the bases of big trees to kill them, or they were girdled with stone axes (girdling destroys the cambium—the vital diaphanous circulatory tissue between bark and wood—and is lethal). The dying would take a while, but the point was to stop the foliage from shading the ground so that the planters could grow their crops. Indian agriculture routinely depleted soils. However, Indian plots were fairly modest. The habit was to move the village when the crop yields declined so lands farmed had a chance to regenerate. Indians did some burning to clear out the understory plants, to be able to see distant prey, or enemies, or to drive game.

In time the forest would return, but fire and cultivation would have changed the terms of the conversation among the different tree species. In Colonial Point, the fire suppressed the sugar maples and beeches. Sugar maple is intolerant of fire and beech seedlings don't like a lot of light. In addition to killing off seedlings, burning can destroy the maple and beech seed bank, making recolonization by beech especially unlikely. (Winged maple seeds—samaras—can fly in from a nearby stand, but heavy beechnuts must "walk," carried and buried by mammals. Once they are established, however, beech seedlings may predominate because they aren't tasty—browsing deer prefer maple and other seedlings.)

Meanwhile, the fire-tolerant oak seedlings just hunkered down to await their moment in the sun, and got it. In the early twentieth century, after the American invasion, Colonial Point was selectively logged for beech and maple which left the oaks to continue. So now the oaks are hundreds of years old, with graceful leans and branching crowns. They cannot live forever, though, and it presents an interesting conundrum. Today's Colonial Point forest is anthropogenic, not an inviolate grove. Unless the land managers decide to burn, these oaks will be succeeded by maples and beeches, and the Colonial Point forest, in its present manifestation, will pass away.[11]

Fascinating as we found the forest history, the object of the field trip was to get some tangible sense of the forest primeval, to con a whole world of it. Rachael and I visited on an April day, some time before the trees leafed out. There was no foliage to impede the pale sunshine from streaming onto the taupe collage of leaves that had been pressed against the earth by vanished snows.

After a brisk chatty gallop through the preserve guided by the conservancy educator and field biologist, who had to return to their nearby offices, we returned unaccompanied to the woods for some quiet time. In truth, we both longed to see the unlikely bear, a visitation to link this present with a primordial wildness. As we proceeded down the dirt track that bisects the 300-acre woods, we saw in the far distance some very big somethings crossing the road, somethings bulky and, we thought, very un-deerlike, the answer to these maidens' prayers. Doubtful that these were bears, but we couldn't say absolutely that they weren't.

After we'd walked a little farther, we left the road and made our way 100 yards into the woods. The ground in such woods is irregular, gently figured, faintly "hummocky." Time, weather, and reducers all have their way with fallen trees; the trunks decompose into soil, or are overgrown by other trees, such as yellow birch, and the upended root masses become tip-up mounds. The ongoing result is a surface that rolls and pitches and tilts the walker. It gives many occasions to be grateful for the range of motion afforded by ankles as they seek balance and accommodate the organic ungeometries beneath. As a suburban child, I grew up in a motley savanna where tree species brought from the four corners of the Earth shaded our neighborhood yards. The understory was sterile as a crew cut, and the herbs and flowers were severely regimented in their beds, expected to blossom loudly, and on time. It was Nature beaten into a tawdry submission, better than no Nature at all, but an unlikely situation for developing the kinesthetics required to move through a real woods.

Irregularity and nonmonotony are the essence of the natural world: no flatness, no straightness, no ninety-degree angles. One must tread more sensitively in such places. It is somewhat a matter of survival. You can't go heedlessly charging forward because you might catch your toe and take a fall, or twist your ankle, or step on a branch and whack your face. This is good. To relocate a part of your awareness to the soles of your feet gets it out of its ghetto in the cranium.

Rachael and I found our spot, spread a blanket on the ground, and just sat still. Metal fenceposts daubed with paint denoted transects of the woods, survey lines laid out by University of Michigan researchers in scientific pursuits determining the frequency of the occurrence of certain species, observing the appearance of an ephemeral stream; documenting other changes occurring over time. Like bands on a bird's leg, these markers were a jarring intrusion on the wilderness fantasy. Necessary, perhaps.

The woods were bare but not stark. The day was mild, the sunlight vernal, and spring was close at hand. On our walk we had seen the first spring beauties, delicate lavender stars dancing on a threadlike stem, and the frilly foliage of Dutchman's-breeches, but no blossoms. Tiny things, vivid at our feet, acquiring a great significance. There is no border between the bottom layer of leaves and the beginning of the humus, but a continuum of matter changing its form. Dig down through increasingly dense tiers of leaf litter and you find the sepia-dark soil, and it is vital, alive with minute stirrings of beings: meiofauna. "Their vigorous and prosperous communities," writes Rutherford Platt, "include: gnats, mites, ants, ticks, maggots, springtails, thrips, spiders, earwigs, mealy bugs, millipedes, centipedes, sow bugs, and tardigrades." (Tardigrades, as their name hints, are slow-moving little arthropods.)[12] Rachael picked out a lacy skeleton of a leaf from the litter, its veins exposed by the chewing myriads. She teased it gently and said, "I had no idea that they were so strong."

Veins of white pine occur in hardwood forests, casting their shade all year round, raking the wind with their needles—*soughing* (pronounced sou or suf, this verb is mandatory in describing the sound of pines). In his book on Michigan's history, Bruce Catton writes about the experienced land-lookers of the era of the Big Cut, timber-cruisers who sought the white pine and could locate pine stands by their characteristic sound. Catton talks of these pines whispering "suicidally" in the breezes, betraying themselves to the loggers.[13] White pine was prized for its clearness and lightness—it could be floated to a mill, was strong and straight, built the cities of the Midwest, the houses of homesteaders on the plains. White pines and their associated red pines could grow to enormous size (well over 100 feet tall and 4 feet in diameter) in pure stands, although not anymore. The great groves of white pines are long gone. A museum-quality stand of forty-nine acres exists at the Hartwick Pines State Park, about seventy miles from here, and they are soaring, sombre, even baleful, but their status as forlorn relics couldn't be more blatant.

Because the white pines were famously the timber tree extracted from this beautiful peninsula, and because the current monocultural tree stands on cut-over lands are conifers, although seldom white pine, when I first came to Michigan I had the vague idea that it once had been one solid stand of white pines. Even though I was at that point living on land with patches of hardwoods, I had to read to believe that the deciduous forest was the climax community in this place. But the mystique of white pine, whose craggy majesty is emblematic of the skyline of the North Woods, does not abate.

Before I came to Michigan, it mainly signified Detroit, and a dim recognition that my father's parents were from here. Harold Francis Mills and Helen Sedina Beacom were from Houghton and Marquette, respectively, towns in the Upper Peninsula. The U.P. shares some qualities with the northwest Lower Peninsula and was inhabited by the same groups of Indians, but it's got different weather and a different history and it means something quite different to Michiganians. The cumbersome designation we use down here—"northwest Lower Peninsula"—amuses people from outstate, but the specificity is important.

In April 1992, I got to make my first real visit to the Upper Peninsula. I gave a talk in Marquette and afterward went on to Houghton, where the Millses had lived and where my father's father learned how to be a mining engineer. It was the hub of a copper and iron mining district, at the hilt of the Keweenaw Peninsula, which juts into Lake Superior, so huge and wild and strange. The Upper Peninsula is another country, a resource colony in the far north where January temperatures may dip to thirty degrees below and annual snowfalls average 160 inches. To exploit the mineral and timber wealth that was there, it took a lot of fortitude, not all of it homegrown. Finns, Norwegians, Swedes, Italians, Cornish folk, even Lebanese came to this far end of the Earth to labor in the mines or engage in the commerce in the little mining towns. Upright frame houses line the residential areas, imposing red sandstone facades front on Houghton's main street. Dairy farms checker the outlying areas and renegade apple trees from long-abandoned homesteads have infiltrated the woods.

These days enough has been improved about clothing and housing that people can live in this region in reasonable comfort, but during the Victorian century it had to have been brutal enough for the feeling of being at war with the elements to be pronounced. It was another place to shoulder the white man's burden, another place to corset desires and override pain while cultivating the virtues associated with dominance. Nature was commandeered to requite man's technological ambition, and there was an unwarranted optimism that it endlessly would. This ingrained a flock of bad habits, the concept of land as a commodity underlying them all. By this reckoning, standing timber is a resource going to waste. Nevertheless, a couple of midwestern magnates had the foresight and generosity to donate extensive Upper Michigan wildlands to the federal government for posterity. One of these areas, Sylvania, is well known and heavily used, the McCormick Wilderness considerably less. These two tracts contained all the Great Lakes Forest old-growth I'd ever heard of.

On that April visit to Marquette, druidic happenstance disclosed some private forest more primeval. Following my talk, a couple of backcountry rangers, ecologist Doug Cornett and photographer Dan Wells, came up to tell me about virgin forest remnants they were discovering in an area nearby called the Michigamme Highlands. They spoke of ancient groves of white pine, white cedars pushing the half-millennium mark, sugar maples several centuries old, all tucked away in the jumbled nubbins in one of Earth's oldest mountain ranges—the rocks are two billion years old. These vestige forests linger in areas too rugged to log profitably—yet—and occupy land currently for sale or in the possession of paper companies. There was a little high-grading for white pine some years ago, and, more recently, helicopter logging for bird's-eye maple, which is much sought for veneer. Doug and Dan and fellow activists Kraig Klungness and Catherine Andrews were beginning to shape a campaign to preserve these ancient forest tracts and weave them, along with other relatively undisturbed state and federal lands, into a grand wilderness reserve in the Upper Peninsula.[14] I expressed a willingness to help and a very keen interest in seeing these incredible places. So a field trip was arranged for August 1992.

My stereotype of virgin forest had mostly to do with the stupendousness of the trees. It turns out that the virginal quality lies not in grandeur, but in refraction. The highlands harboring these remnants are nowhere level but are a tumult of defiles, slopes, faces, wetlands, groves, and pockets. It makes an uncommonly rich array of habitats, where every cubit is a world of its own.

A raven croaked greetings (or possibly "Beat it!") as we approached the rock cleft up which our party made its entry into the highlands, and their revelation of ecological integrity. We hiked through dim, pure stands of old hemlock, past exquisite little iris bogs caught in depressions in the broken terrain, past sphagnum openings with palisades of white cedars, up to a promontory looking out over lakes with rocky, spruce-clad islands. In the canopy were occasional crimson sprays of sugar maple, already flaunting fall color. There were waterfalls and seeps, hewn basalt faces washed with moss, Indian pipe standing scarce and nacreous. No one yet has identified all the trees, shrubs, grasses, sedges, forbs, fungae, mosses, and lichens enmeshed in that wild redoubt. Indeed such cataloguing may be beyond possibility. This area contains a diversity of mite species believed to be equal to that of the tropical rain forest. An abundance of wildflowers lay underfoot: the mellifluously named pipsissewa, snowberry, bunchberry, rattlesnake plantain, round-leafed orchis. We gobbled dewberries, blueberries, and huckleberries. Although we were loping, trying to see as much of the territory as possible during the long hours of summer daylight, in the moments when we paused, everywhere the eye alighted was a new constellation of living forms.

Even within species there was stunning variety. Trees of all ages were present, from yearling white pine feathers to soaring monarchs three armspans around; from elfin sugar maples fluttering their jaunty leaves to craggy old geezers whose trunks ascended out of view. There was a procession of generations, not orderly, but complete. Elsewhere in this region, where the topography is more forgiving, logging and pulp cutting have opened edges and acres of browse. White-tailed deer have irrupted and pressed hungrily into deeper woods, eating away the cedar and hemlock seedlings and tender plants of the forest floor. Where we were, only rogue bucks could survive the winters. A misstep into the shoulder-deep snow could make them easy game for the timber wolf and cougar whose presence here is surmised from tracks. Beautiful as the place was, its extraordinariness did not register until after we'd departed, and were driving past endless pulp plantations of jack and red pine and high-tech larch. In some primal part of my being, I had recognized the virgin forest. Its wholeness was somehow familiar and the organism that I am felt quite at home.

Friends on the hike took photographs of us standing in a grove of virgin white pines. Later that year I carried those pictures with me on my trip through India. I would consult them to remind myself of the splendors of the North Woods; of the immensity of those pearly trunks dwarfing any four of us in breadth and reaching up past maple, hemlock, and birch to poke their windswept crowns up over the canopy. Compulsively I showed these pictures to the people I met on my travels, wanting to share some sense of astonishment that such trees could exist, anywhere, ever, and yet.

The Michigamme Highlands is rugged and rocky. Its topography and soils are quite different from those where I live, so it isn't fair to extrapolate the primeval look of the land I am standing on from a woodland such a distance away. Yellow birch would be more common north of the straits of Mackinac than it is in my home forest, and beech less prevalent. On the Leelanau Peninsula it was easy to log; in the Michigamme Highlands, very difficult. Hence no extensive ancient forest remnants anywhere near me, certainly no tracts large enough to accommodate a full day's walking. The impressions gathered from reading, from hiking that U.P. forest, and strolling the woods at Colonial Point, must suffice to compose a vision of what to restore to; a vision of what Nature deemed optimum for this place.

Maybe it's about charismatic macroflora. One hears, in conservation biology, of charismatic megafauna—lions and tigers and bears, photogenic wildlife that excite empathy, and preservation campaigns. One also hears of the less conspicuous, but no less essential, enigmatic microflora—like the many kinds of mycorrhizae, the soil fungae that link the root hairs of different members of plant communities, be they of forest or prairie, and help the flowering plants draw their sustenance from the soil.

Could it be that awe is a necessary ingredient in psychic health, and one sure way to feel it is to be dwarfed by a living creature? If so, then the charisma of trees in ancient forests becomes quite understandable. To envision what vast populations of them must have been like is a pastime for the armchair explorer. Desperately scarce now, it's incredible to think that such forest was regarded as a damned impediment to agriculture, something to be done away with by girdling and burning, chopping and sawing.

Chapter 3

Woods, Woods, and Nothing but Woods

From my studio I can gaze out across an old field to a tuft of hardwoods beyond, these covering just a fraction of this square-mile section of Kasson Township. What remains seems fairly beat-up and washed out in comparison with a woods that would have been prowled by wolves, bears, and cougars, woods that still had top carnivores and great populations of other kinds of creatures, too. Woods like that were right here, and not so very long ago. Joseph Krubner, a settler in North Unity, an early Czech and German village a few miles due north of here, wrote of his experience in the winter of 1855: "For a while hungry wolf's (*sic*) were chased away from our doors. But with approaching spring, when the snow melted and the lake still frozen, no boats were able to reach us, potatoes and what ever we had was gone, hunger begins to strike again. By the time it reached its peak, we were saved as a flock of wild pigeons came by. Everyone who had a gun and was able to use it was shooting them. Few of small lakes helped to change our menu as they were full of fish."[1] The same forests that gave refuge to the varmints sustained the pigeons and even the watershed whose above- and below-ground tricklings made for lakes and fish.

The forests could sustain all of that life but not a culture that regarded trees as a resource at best, a nuisance at worst. In 1880 Js. M. Neasmith provided this description of Kasson Township, on the southern border of Leelanau County, to the Michigan State Land Office: "land surface generally rolling—soil varies from sand to sandy loam bottomed on clay and mixed with lime and coarse pebbles—soft and spongy—Principal timber sugar [Maple] and beech with Elm, Ash, and Lynn [Basswood] and on the ridges hemlock."

At one point in my researches into the history of my own land here, as I attempted to trace the parcel back to its original patent, I came across a record of a turn-of-the-century timber sale of the north half of the northeast quarter of the section. (A section, 640 acres, is one-thirty-sixth of a township, which in the Northeast and Midwest is both a geographic unit and a unit of local government. Counties are comprised of townships.) I'm sure the sale seemed then a practical and necessary thing to do, just as it does to my neighbors a half a mile away who have been selectively logging their woods. It's not uncommon to see logging trucks on the back roads of the county, some of them hauling substantial tree carcasses.

Some of this writing is about trying to understand that tree-killing habit, and to understand why this culture's been so uncherishing of our great old trees, and the ecosystems for which they stand. Most Old World forests were doomed from the dawn of cities, smelters, and navies. These days, the greatest conservation battles rage over ancient forests in the tropics and North America. Only oceans and great mountain ranges purchased survival into the modern era for the forests of North and South America. Once the oceans were crossed, and ways up the mountains were found, the New World forests began to fall.

To begin a true and lasting restoration it is necessary to travel imaginatively back through time to earlier, less disturbed landscapes to develop an authentic vision of what to restore to. How else can we view the losses, the irreparable gaps rent in the fabric by extinctions, but as remonstrance? Santayana's dictum that those who cannot remember the past are condemned to repeat it is germane. More than just recollecting the past—our patterns of using land and their consequences—it is necessary to show a little remorse for the damage done and to trouble ourselves with the question, Why?

Environmental history is the discipline that will provide us with the memory, the remedy for the cultural amnesia which leaves us with only a generalized and hazy sense of landscape, and a numb acceptance of the stripped remnants that exist. My own interest is to see the practice of restoration bedded in an understanding of environmental history, the more local the better. An environmental history is the point of departure for ecological restoration, the place's story of past and present yielding wisdom for its future.

I take up this history of my home place because it concerns the fate of the land, the most important story I can know. The histories of our life-places may inform our visions of the good life we might wish for our natural communities, and for their families of beings, human and nonhuman, long into the future. In the United States, which has such a very brief history as a nation, understanding the velocity of the radical transformations (and simplifications) of our landscapes can teach us that all manner of "limitless bounty" can be exhausted. Our land-use histories are usually catalogues of mistakes not to repeat. Maybe the biggest one is underestimating the ultimate impacts of incremental changes.

The knowledge that the rich and graceful hardwood forest which for several millennia clothed this bioregion could be felled within less than a century shows that the spirit of the local pioneers was hostile to the local gods. The myth of the limitless frontier overpowered common sense and simple observation. Each successive cutting, and the increasingly extensive clearing of the woods, has exposed more soil to erosion and diminished the land's richness.

49

Anybody with the least interest in American natural history will know something of the passenger pigeon, of its staggering abundance, perhaps from Audubon's accounts of sitting and watching solid flocks pass overhead for days. Those stories have a mythic quality. Closer to home the pigeon's extinction became more tangible. In Leelanau County there were great roosts of passenger pigeons as recently as 1877, recounted thus by William Joseph Thomas, an early Northport resident:

> The first twelve years we were here, from 1856 to 1868, we used to see vast numbers of wild pigeons all through the warm season of the year, from early spring till the first snow in the fall. I do not know where they wintered but presume they went south to Texas. It was a grand sight in the spring of the year just as the snow was disappearing and the sun was bright and warm to go out about nine o'clock in the morning and watch the pigeons flying. Their first appearance in the spring would not be large. But in a few days the flight would begin at sunrise, and soon the whole skyline would be alive with pigeons flying as close together as they could fly, and possibly for a minute or two there would be no openings in the sky at all. This would keep up until near noon. The next day would be the same scene over again. These flights would continue for one or two weeks every spring. Hundreds of thousands of pigeons would stay in our county all summer, feeding on all the kinds of nuts and berries as they could find. The first few years that we were here, the pigeons were only hunted and killed for home consumption, but later they had a market value, and a great many people engaged in catching them.[2]

They were sold by the crate, six dozen for a dollar, or in uncounted quantities by the barrel. Wrote Aldo Leopold, "The pigeon was a biological storm. He was the lightning that played between two opposing potentials of intolerable intensity: the fat of the land and the oxygen of the air. Yearly the feathered tempest roared up, down, and across the continent, sucking up the laden fruits of forest and prairie, burning them up in a travelling blast of life. Like any other chain reaction, the pigeon could survive no diminution of his own furious intensity. When the pigeoners subtracted from his numbers, and the pioneers chopped gaps in the continuity of his fuel, his flame guttered out with hardly a sputter or even a wisp of smoke."[3]

About a century after the first Europeans crossed the North Atlantic to begin to explore the continent of North America, the French arrived and began moving westward through the St. Lawrence watershed to the interior province of the Great Lakes. Around 1634 Jean Nicolet, a French explorer, was out West in Michigan at Champlain's behest, hoping to find Peking.[4] There were outposts in Michigan to the northeast, at Sault Ste. Marie, by 1671, to the south at Detroit, from 1701, and at Fort Michilimackinac on the Mackinac Strait in 1715.

The now-industrial southern Lower Peninsula of Michigan is "old," having been settled a century and a half before the northern region. Southeastern Michigan was a site of early French military bases, and an embarkation point overland for settlers from New York and New England who began arriving as soon as the Erie Canal and Indian treaties allowed them access to what was then the Northwest Territory. Their frontier was by turns swampy and sublime, impassable mud in some places and parklike oak openings in others. Some settlers came by land from the south, from Ohio and Indiana, through an immense mucky old lake basin. The miseries of such overland travel may have given them plenty of reason to despise what we now call wetlands, and to regard drainage, like clearing, as a sure way to improve upon Nature.

Southern Lower Michigan, bosky in spots, was by and large a swampy, frigid, forbidding place. The weather was and is grim, there's not much sun and a fair amount of snow. It must have seemed terribly gloomy to the few settlers who filtered into its interior during the early nineteenth century. One thing it did have, though, was trees. In the Saginaw Valley industrial lumbering had begun by the 1830s. This was a landscape-wide and immensely profitable rendition of the homesteaders' bitter struggles to clear their acreage for farming. Once de-

51

forested and drained the Saginaw Valley became superb agricultural country. The lumber economy took off, and in little more than fifty years virtually all of the great stands of white and red pine in Lower Michigan had been logged, and much of the pinelands incinerated. Sometimes the fires were set deliberately, to "prepare" the soil, sometimes they were accidental, caused by errant sparks from passing locomotives.

In a 1929 U.S. Department of Agriculture bulletin, "The Economic Aspects of Forest Destruction in Northern Michigan," William Sparhawk and Warren Brush wrote, "Of Michigan's original stand of 380 billion board feet of saw timber approximately 35 billion feet was cut and burned in clearing land; 73 billion feet was burned and wasted during or after lumbering or destroyed by forest fires independent of lumbering operations; 204 billion feet was cut for lumber; and 40 billion feet was cut for other products. . . . In many parts of the state, the amount of timber destroyed by fire exceeded the amount cut. . . . An extremely conservative estimate of the aggregate value at the mill, at the time of cutting, of all the sawed lumber cut from Michigan forests is $2,500,000,000. . . . If the value of other timber products be added (logs, poles, posts, ties, shingles, staves, etc.), the total of over $3,000,000,000 will be ten times the value of all the gold that has been taken from Alaska, and more than twice the value of all the gold produced in California."[5]

Northwest Lower Michigan—my bioregion—has, or had, a different forest (maple-beech-birch-hemlock) from south and central Michigan. The region has a different history as well. The earliest settlement in the 1850s came from the northwest, by water, and Protestant rather than Catholic missionaries played an early decisive role in the post-contact destiny of local Indian groups. But here, too, the woods went fast.

Despite the occasional conflagration of logging slash, in Leelanau County subsistence farming became the rule. Whereas earlier Michiganians had come from New York State and western New England, a second wave of settlers came from Germany, Britain, Poland, and Czechoslovakia by way of the great industrial cities—Detroit, Michigan; Gary, Indiana; and Milwaukee, Wisconsin—hundreds of miles to the south. As soon as the lands deforested in the first cut and left uncultivated began to regain their green cover—say, within thirty years—resorts were established on the shores of Lake Michigan, Lake Leelanau, and along the west arm of Grand Traverse Bay, which cleaves the Leelanau Peninsula from the northwest quarter of Lower Michigan. Summer visitors have long been another cash crop.

In the first half of the twentieth century, the uncertainties of the U.S. economy had people traveling the length of the state from factory to farm and back again, seeking cash and subsistence by turns. As the county soils forgave less and quit producing, and as the U.S. slid into the Great Depression, farmers were forced to abandon their lands for nonpayment of taxes. The state of Michigan picked them up and began developing the state forests, many of them monocultural stands of pine. The Civilian Conservation Corps also planted many millions of trees, creating national forests (none of which are in this vicinity). World War II unleashed technological and political changes that ripped through rural communities like tornadoes. If the landscape of the North Woods has been revolutionized, so has its human society, and these upheavals in land and life have left some physical and cultural impediments to the attainment of true community. The human community now living here resembles the land's geology—groups have been moved and tumbled and shaped, mixed together and then deposited on the surface by forces far greater than they—economics, history, *Homo sapiens'* immemorial restlessness. Nowadays there are people scattered throughout the region who are again claiming home ground. They are new settlers, intent to make soil—immediately, in their compost heaps, and in the larger, longer sense of allying themselves with and defending the whole complex of life forms and streamflows that arise from, nourish, adorn and define this reach of the Earth.

Susie Stachnik, née Skipski, was the youngest of the nine Skipski children born in the farmhouse that until recently stood next door. I lived in that house when I first came to Michigan. I visited Mrs.

Stachnik, thinking perhaps she might be able to tell me what this land looked like when she was a child; that I might get some image of the earlier landscape from her, some account of what had been done to, or with it. Like the early settlers' families around here, such questions can generate a lot of offspring. Mrs. Stachnik possesses a friendly dignity and winsome midwestern reserve. In her late sixties, she is a pleasant person, attractive and trim. She invited me in to her immaculate house, and we sat together on the living room couch. A composite portrait of her parents, Joseph Skipski and Mary Skipski (née Fleis), hung on the wall. Mrs. Stachnik told how an itinerant photographer had come to the farmhouse and taken away the separate photographs of each of her parents magically to combine them in a handsome frame for the price of ten dollars, which then had "seemed like a lot of money."

The farm's eighty acres, the west half of the northeast quarter of the section, came into the Skipskis' possession in 1909, although they may have been in residence there before then. Susie's sister Betty pegs the date of construction of the barn, still standing, as being 1910, the year of Betty's birth.

The Skipski family lived on the farm for almost half a century, an atypically long time as things went in the rural U.S. Their household was largely self-reliant. The Skipskis grew most of their food—meat, milk, fruit, vegetables, and grain—and got much of their fuel and building material from their land. They raised potatoes as a cash crop for money to purchase tools and cloth, and the occasional treat, like an orange in midwinter to lift the spirits of a kid with a cold. The children evidently went their separate ways because by mid-century, when the parents had got too old to farm, there was no one remaining to take care of the land and the family.

Susie Stachnik had been a hardworking farm girl. For reasons that became obvious as she described her family's life on the farm, I realized that my romantic inquiries about the woods and ponds out back were simply irrelevant to her experience. Because she was a contributing member of an every-waking-hour enterprise, she had not had the opportunity to acquaint herself with the natural history of her immediate surroundings.

Mrs. Stachnik related a life demanding beyond my comprehension. In 1986 I found living in her birthplace crowded enough with only five other people (three adults and two children), and, thanks to our roommate-landlord's penchant for building additions, we had a great deal more floor space than the Skipskis had. By my day, we also had electric lights and indoor plumbing, which had not always been the case in that farmhouse. Trying to imagine what it might have been like in, say, the dead of winter for a family of eleven back in 1925 was mind-boggling. The character of relationships in those times seems to have been so different. Mrs. Stachnik said she didn't know if her mother really liked her father or not. "It's odd how they picked their mates," she remarked, referring to the custom of fathers marrying off their daughters. Her mother, who came from around Isadore, about six miles north-northeast as the crow flies, said "she was real lonesome." Years would pass between visits with the mother's family. All the transportation they had in those days was horse-drawn.

The family farmed everything, she said. There was a garden in which Mrs. Skipski "grew everything that we liked to eat. We raised our own chicken and pigs, made all our own sausage and headcheese. Evidently it had to have been colder . . . we had a nice granary where we'd keep sausage rings on poles." The climate was such that these meats kept through much of the fall and winter. The Skipskis put salt pork in barrels and canned other meat. The six cows that Susie milked every morning and every night produced butter and cream that was kept in the farmhouse basement. They grew rye, wheat, and oats, which they fed to their horses, and corn. Every year a local thrasher brought his big machine by tractor. The grain was threshed on the barn floor.

"We grew beans," said Mrs. Stachnik. "My father made a machine for sorting beans." On winter nights the children would vie to operate the machine, evidently a true labor-saving device. Mrs. Stachnik mentioned that in the winter, too, "my mother would spin wool.

She'd get a bag of this wool and we would wash it and we'd card it and she'd spin it." Her mother knit constantly, "knitted all these wool socks and mittens."

After trying to elicit Mrs. Stachnik's recollection of the landscape of her childhood, thinking that she would have memories of trees or scenes, or specifics, it became clear that land health was, to her, defined in terms of cleared fields, propriety of the farmstead, and its good repair. More to the point, northern Michigan farm girls of the thirties simply had no time for nature study, what with milking cows and spraying potatoes and walking up and down the rows of grain, hand-weeding. It was a hard life, but Mrs. Stachnik said she thought it must be preferable to what kids have to deal with these days.

Michigan was part of the Federal Land Survey ordained in 1785 by the newly constituted U.S. government's General Land Office. To facilitate the settlement of the U.S. from the "point of beginning," where the Ohio River crosses the western border of Pennsylvania, the government embarked on an amazing project: a mile-by-mile survey of the North American continent which eventually progressed west to the Aleutian Islands and south to the Florida Keys.

Surveyors visited (or claimed to have visited) every square mile of the old Northwest Territory, measuring with their 66-foot Gunter's chains, noting landmarks as points of reference for their descriptions, cataloguing 640-acre sections for homesteading and township government use and, in the process, recording the face of the landscape—vegetation, soils, streams—as it existed before the extensive disturbance brought on by settlement. All of this information was published through government land offices, and for a dollar an acre, the prospective pioneer could choose a parcel (a 160-acre quarter-section, for instance), sometimes sight unseen.

Eventually hundreds of millions of acres were transferred out of the public domain by this process. It was a laying down of straight lines on the land, the establishment of a grid of mile-square sections grouped in townships of thirty-six square miles. It's testimony to the newness of the United States and to its gestation in the Enlightenment, as well as to the rolling real estate boom that has been the demon of much of American history, that we possess this detailed inventory of *terra incognita*, a square-mile by square-mile description of much of this continent between Canada and Mexico.

Surveyors were directed to go out and subdivide the country, and to leave their marks on "witness" trees or, in the prairie provinces, to place stones or drive stakes to denote the corners of the sections and the center points, or quarters. The surveyors had to choose "bearing" trees standing near the spots they needed to denote, avoiding species like white or red pine that would likely be logged. It's fairly ironic that the surveyors, by whose labors the ecosystems of much of America were reduced to private property, were at the same stroke providing one of the most systematic portrayals of an uncharted wilderness that had ever existed, an invaluable record obtained by human observers traveling on foot, rationalized by geometry and the compass.[6]

Kasson Freeman, after whom Kasson Township is named, was born in Chautauqua County, New York, in 1815. He was in the tailoring business in Wisconsin until 1863, when he moved to Leelanau County to homestead. A desire to grow fruit, as had been the way in Chautauqua County, said Freeman, along with the presence in Leelanau County of some friends, suggested the move eastward. Freeman first visited his Michigan parcels in December 1861, having crossed Lake Michigan by steamer and then "loaded our things on the sled and started for the woods, woods, and nothing but woods." Freeman made a reconnaissance and the beginning of a homestead—a cabin of thirty-two-foot-long "house logs." His wife, Lodenia Salina Ward Freeman, came to join him, with their children, in 1863. Lodenia Freeman was a doctor, in her time, the only one in this part of the county. Once settled, Kasson Freeman transported his fruit trees from his old home and shortly thereafter began impressing fair-goers in Traverse City, the nearest big town, with his produce.

A string of disasters befell him, from the loss of his hearing to the loss of his fruit trees to the death of his wife. These griefs drove him to Louisiana where he stayed until 1893. Then he "began to feel a desire to see my old friends and neighbors in Leelanau County. . . .

I was greatly surprised on Reaching this Grand Traverse Region to see the change that had taken place since I lived here. Farms cleared up with nice modern built houses, large framed barns, fine stock with beautiful horse teams in the place of the old Buck and Bright oxen, and the fields as smooth and green as a billiard table, not a stump or root in sight."

A few years after his return to the Leelanau Peninsula, Freeman told his history at one of the first of the Old Settlers' Picnics, an annual event that continues to this day:

It was the free homestead law that induced me to come here, and, I believe, a large majority of all others that came here in the 60s.

In 1861 our generous Uncle Samuel said to his nephews and nieces: "I have thousands upon thousands of acres of unoccupied (*sic*) land, and I would like to see all my children with a home of their own. Now I will tell you what I will do. If you that have reached your majority will go and select a quarter section of 160 acres of this land, move on to it and make it your home for five successive years, I will at that time issue a patent to you for it, and then you will have a home of your own choice and selection."

Now, was this not a grand offer? Was it not timely and well worded? Did he say, "You farmers"? No, it took in everybody and was a glorious triumph for women's rights. No law was ever made before or since we Yankees twisted the Lion's tail until he let go the grip on our throats, and with a roar skipped to his den across the Atlantic, that has done for our citizens, both male and female, what this law has. It has made farmers for five years at least of old maids, chambermaids, milliners and schoolmarms; old men, young men, tailors, shoemakers, lawyers, doctors and preachers. All citizens of the United States, 21 years old (except married women living with their husbands) are entitled to the benefit of this free homestead law be he or she black, white, skyblue or scarlet, as there was no color line drawn.[7]

In addition to peopling the land and transforming it rapidly, one of the effects of the new geometric survey method was to impose a habit of rectilinearity on the rural landscape. Other cultures, even the original North American colonists, used metes and bounds, more top-ographic definitions of their parcels, which often determined property lines according to landforms or watercourses. In older societies these realities were at least conceded, whereas in the territory of the straight line, it sometimes seems like the elimination of the third dimension—of vegetation, of relief—is the ultimate agenda. Undulating terrain is inimical to machinery. Thus the level undeviation of our fields, our forests, and certainly of our roads, except the roads that follow the traces of Indian trails.

So I'll make no bones about it: one reason restoration is impor-tant is that it may provide a way to turn aside the march of civilization, and to obliterate all the straight lines that empire has loved so well, from the days of the Romans onward. A straight line is indeed the shortest distance between two points. More often than not, the points open gashes and the lifeblood of the land hemorrhages away, some-times in the form of soil, or the species become extinct or extirpated as a result of the fragmentation and reduction of territory caused by the straight roads and fences running between those points.

Prerequisite to imposing a grid on the old Northwest and rendering it salable was the removal, or subjugation, of its aboriginal inhabitants. According to historian Bob Doherty, "surveying re-quired getting title to the land and that meant buying it from the In-dians which imposed a new system of ownership (fee simple)."[8] The Odawa people were more or less minding their own business here, just coming to hunt, then establishing seasonal villages for hunting, fish-ing, growing squash, beans, and corn, and collecting maple sap. The entire Indian population of the Grand Traverse region from the mid-eighteenth to the mid-nineteenth century was perhaps 750 souls. It was a transitional era, between the waning of the fur trade, in which the Odawa were briskly engaged as go-betweens, and the onslaught of American settlement.

By the mid-nineteenth century, when the U.S. government sought to treat with them, the Michigan Indians had been dealing with foreigners, their technologies, and their diseases for nearly two centuries. The Odawa had been in the habit of trade and negotiation for a century or better and negotiated brilliantly, trying to make the best of a bad situation. They were not unaware of the fate of other groups of Indians, and they succeeded in winning the right to remain in Michigan; not to be relocated to the West, along another trail of tears. What would not be allowed the Indian people was the right to hold their land as a commons. As part of the federal policy to detribalize the Indians, the U.S. government required that they select parcels of land which would be deeded to them as individuals or heads of household. Thus the Odawa and Ojibway were estranged from their habitat by treaty provisions that required them to own and farm land individually—"in fee severalty." This imposition of the practice of treating land as property that could be privately owned was a violent departure from their customary allocation of hunting and fishing rights to family groups by usufruct. The tracts of land allotted to the Indians were far too small to support them, and as a result of this incomprehensible identification of property rights with a piece of paper, and the monetization of economic activity, many of the Odawa people were, within a couple of generations, deeply impoverished.[9]

The near-complete displacement of the Odawa and Ojibway bands living on the Leelanau Peninsula from their lands took only about twenty years following the Treaty of 1855, which had reserved thousands of acres of the northern portion of Leelanau County for allotment to individuals and families. Thanks to American settlers' claim-jumping and to bureaucratic ineptitudes, less than a quarter of the reserved land wound up in Indian possession. By 1880 about two-thirds of these allotted lands had been sold to white landowners (in part through the agency of two opportunistic band members) or were confiscated for delinquent taxes.[10] Today all the land base that the Grand Traverse Band of Ottawa and Chippewa Indians retains of the better part of the five Leelanau County townships set aside before 1855 for their exclusive use are 150 acres of tax-reverted lands (conveyed by the state to the county in 1943 for "Indian community purposes") and some recently purchased tracts totaling approximately 250 acres.

The descendants of the native people who preceded the American settlers have struggled to achieve for themselves first citizenship, and then the benefits that the U.S. government confers upon an officially recognized tribe. A cluster of houses, a motel, a Catholic church, and two casinos constitute Peshawbestown, which lies twenty miles northeast of my place. A settlement begun in 1856, Peshawbestown is where many of the 420 local members of the Grand Traverse Band of Ottawa and Chippewa Indians now dwell. The mainstays of the band's economy are a casino, which the tourists to our area like, and a small fishing industry, which so-called sportsmen who come to this area do not. The Grand Traverse Band have an ongoing fight to maintain their treaty fishing rights. Indian fishermen are usually vindicated by the courts and excoriated by public opinion. Nevertheless, a small handful of Indians are able to continue their long practice of getting some of their living by fishing the surrounding waters. Far more of them, however, are employed in the gambling industry and in the administration and programs of the Grand Traverse Band.

No eighteenth-century Indians pursuing their subsistence from Nature live here anymore, and even the earlier groups of aboriginal people who lived in this region were constantly adapting to changes in circumstance, and adopting new ways. But lacking the means, and possibly the inclination, the native population didn't engage in wholesale conversions of its landscapes. That awaited the commoditization of land itself.

The animus of the nineteenth century was surely the steam engine, and to fuel its iron metabolism was the fate of the forest in the northwest part of Lower Michigan. Nearly as I can tell, the lumbering that went on there was discontinuous from the great logging boom 100 miles to the south. The loggers here were not so itinerant, nor were all of the lumber entrepreneurs. The "wooding" industry—supplying fuelwood for the Lake Michigan steamboats, called "propellers"—was a primary occupation here throughout the snowy winters. Tree cutting began on South Manitou Island in the 1840s, moved to the mainland, peaked around the 1880s, and was done by the teens of this century.

One bit of local history affirms the fact that metallurgy has been a prime cause of deforestation since ancient times. In the 1870s in Leland sixteen miles to the north there was an iron smelter situated on what was then called the Carp River, which drains Lake Leelanau, the long body of water that extends nearly the length of the center of the Peninsula. The idea was that the Peninsula forests could be turned into fuelwood, which could be moved from various points throughout the Leelanau Peninsula to the lake and transported on the lake to the foundry, where the wood would be converted to charcoal to smelt the Upper Peninsula iron ores being shipped eighty miles across Lake Michigan from the docks at Escanaba. The availability of markets to the south in Detroit, it was thought, would make such a far-flung enterprise a sure thing, but the foundry only lasted about fifteen years. During its short existence the iron works consumed more than 400 cords of wood a day. A cord is 128 cubic feet of wood, 4' × 4' × 8'. Wood cost about two dollars a cord in those days, and came in four-foot lengths.[11] (Nowadays firewood, which comes in sixteen- to eighteen-inch lengths, costs about thirty-five dollars a cord in Leelanau County. These figures aren't in constant dollars—inflation no doubt accounts for much of the difference, but so does the diminution of the forest.)

The woods were also cut to supply hardwood flooring to the Chicago market, as well as to produce some other, locally finished wooden goods such as maple bowls and shoe pegs. Maple City, capital of Kasson Township, was originally named Peg Town after the factory located there.

Reading about forest history, I am struck by the fact of the phenomenal abundance of wood and trees. The single most vivid fact about this nation, the United States, may be that for the first moments of our history there was wood to burn and wood to build with and wood simply to destroy, to get it out of the way. This immigrant culture couldn't see the forest for the trees, for the forests of their homelands had long since given way to agriculture. These woods, rich enough to provide a small aboriginal population with ample subsistence, couldn't provide the increasingly numerous nineteenth-century

settlers with a European kind of living. Many of the men in these settler families found winter work in logging camps and persisted in
clearing and developing the homesteads they'd return to during spring
and summer. Logging became a way for some of the newly Ameri-
canized aboriginal inhabitants of the area to make a living, and dwell
with their families, in lumber camps in the woods until the woods
were gone. The extraordinary labor it took to clear the trees fairly jus-
tifies the mythic proportions early settlers and loggers have attained
with the passage of their era. The trees hereabouts, judging from the
diameters of the logs photographed on the sleds and railroad cars,
were monstrous, broad as oxen. The practice of nineteenth-century
logging done by hand tools—axes, crosscut saws, peaveys—with
drayage by horses along iced skid roads, was a gargantuan labor, and
extremely dangerous. Hard as it was on the lumberjacks, it didn't in-
flict nearly the damage on the land's capacity for regeneration that industrial logging today does, using infernal machines like feller-
bunchers with their weight and speed and lack of human precision.

As a species we have a tough time considering the long-term consequences of our actions on the land. In all, it took only about seventy
years to log off Leelanau County's 216,000 acres, without benefit of
chain saws. Clearing the land was the obvious first thing to do. It always has been, except in the prairie provinces where fire kept the trees
at bay and where other steel blades—not axes, but ploughs—were
used to subject the wild land to agriculture. More than one local in-
formant, commenting on how quickly northern Michigan's hardwood
forests were logged, has said, "They thought the woods would never
end."

What is the difference between an era and an episode?
Growing up on the standard portrayals of American history, I got an
impression of the westering settlement of the continent as a logical,
orderly sequence, and a period of some duration: landless people from
the old, oppressive country hit these shores, head west, build a cabin,
stay put. The cabin becomes the ancestral home, the cornerstone of
an era culminating in agrarian prosperity. A good many of these Mich-

igan homesteaders, though, had just a quick twenty- or thirty-year sojourn here, one of several in the Northeast, or upper Midwest. No European settlement on these shores has yet attained a life span as long as that of the ancient grove of white cedars somehow still standing on South Manitou Island.

The settlers here who became self-reliant farmers lived in a period that elapsed during a human life span—from the end of the Civil War to the Great Depression. World War II was really the end of that way of life around here, a way of life relatively independent of wage labor, of wholesale commodity production. Stone fruit orchards, cherries especially, do well on the higher, less frost prone ground hereabouts and have been a mainstay of the region's agriculture since the twenties. But such land has of late, alas, become valuable "view property," and is beginning to sprout lavish second homes in lieu of fruit trees. For only a little while—1870 to 1940, say—Kasson Township was a place where people logged and farmed, where kids attended one-room schools, and where social life revolved around the churches or the weekly magic lantern shows in Maple City or the evangelical camp meetings held in the summertime. Not so many years ago, some citizens of Kasson Township assembled a history. The accounts in *Remembering Yesterday* are varied, from dry and thorough genealogy to high-minded sagas of pioneering to droll Twainian tales of earlier times. The photographs, in the main, depict visages sober yet almost fierce in their confrontation of the camera. There is only rarely a glint of humor flashing wife to husband, or brother to sister. The settlers' families were, by our standards, huge. In the stories of early settlers in Kasson Township one commonly encounters families of eight or ten children (Susan Stachnik's grandparents' issue was more than 100 strong by the fourth generation). Divorce was not unheard of. As well as great progeny, there was all manner of human variety—a hermit named Lot, a homeopathic physician, who lies in the cemetery at the end of our road. People who managed to escape childhood diseases like diphtheria, or fatal farming or logging accidents, or mortal complications of childbirth that might cut them down early on often lived into their nineties, judging from the headstones and the obits in the county paper. Other settlers left no trace except the changes in the land.

Only the late-breaking science of conservation biology has begun to articulate the consequence in extinctions of such wholesale land disturbance as deforestation. According to biologist Edward O. Wilson: "Whenever careful studies are made of habitats before and after disturbance, extinctions almost always come to light. The corollary: the great majority of extinctions are never observed. Vast numbers of species are apparently vanishing before they can be discovered and named."[12] Thus during that half-century in which the Leelanau forest floor was bared to light and summer heat and to the unbuffered impact of rain and wind, there was very likely a quantum loss in biological richness, creatures lost and gone forever, for we can never know all that was here in presettlement times. Botanizers and birders, naturalists and chroniclers of Nature, seem to have been underrepresented among settlers in the Leelanau Peninsula.

As I read accounts like Kasson Freeman's, of his homestead in New York and a sojourn in Fond du Lac, Wisconsin, and of a period spent here before his long descent to Louisiana, I realized that the transience of our households, our shallow tenure, is consistent with our history, for better or worse. Legions of these nineteenth-century sojourners vanished unrecorded and are forgotten. Names no longer uttered populate the early plat maps. Old stone basements indent the countryside, great clumps of dooryard lilacs perfume the June air only for passersby; the farmwives to whom they once gave pleasure are vanished as dust. If most of the pioneers are lost to us as individuals, their collective memorial—a pastoral landscape where once there was forest—is what we are left, the signature of a fleeting episode, hardly an era. Yet because of the radical change it wrought in the land, and what the changed land has subsequently made of people, we are as severed from that past and the life it knew as we are from the days of Troy.

A few families have stayed on their lands around here, getting on into the fourth generation, like the Cates over in Solon Township to the east, or the Brights, just over on the next road. Lowell Cate, who is in his sixties, remembers a time when this land was in far sorrier shape. In his lifetime he has witnessed some spontaneous regeneration of its health. Once I exclaimed to him that Leelanau County must have been a lovely place to grow up in, and he said, "Not really—the woods have come back a lot since then."

Chapter 4

Disturbed Ground

Poor land may be rich country, and vice versa. Only economists mistake physical opulence for riches. Country may be rich despite a conspicuous poverty of physical endowment, and its quality may not be apparent at first glance, nor at all times.

—Aldo Leopold, *A Sand County Almanac*

My thirty-five wooded acres of northwest Lower Michigan is by local standards a middling-sized parcel. This almost-sixteenth of Section 26, Kasson Township, Leelanau County, Michigan, is like a foster child, one with post-traumatic stress syndrome.

This latter-day Walden is a poorly edited anthology of the trees of the forests of this and more northerly regions on both sides of the Atlantic. It is a palimpsest of invasions, counterinvasions, and atavisms. On my land, struggling up through the mat of conifers, and tucked away in the little scrap of hardwoods at the rear of the property, are second- and third-growth specimens of the commonest members of this kind of forest: a preponderance of sugar maples and beeches, occasional hemlocks, yellow and white birch, basswood and ironwood, fire cherry and aspen, sumac and ash. For the most part, however, this acreage is a goofy mélange of imported Scotch pines, native jack pines, and blocks of blue spruce and balsam fir.

Even in the best of times, this land is something of a biological desert, notwithstanding the fact that it's got a lot of animal and vegetable matter on it. It's a long thin parcel, a half-mile deep and 660 feet wide at its widest point. In a black-and-white Soil Conservation Service aerial photo of the section, this strip looks like a piece of corduroy, ribbed, with patches worn through. Vivider still, and also courtesy of my federal government, I got a color slide of this section as seen from the air, and had it printed up to serve as an object of contemplation as pinned on my studio wall: the gull's-eye view. It shows this piebald place, *terra mea*, governed by straight lines of conifers; and other parts of the section to the west—Mrs. Rosinski's land— with patches of hardwoods. Her woods curve from northeast to southwest in a handsome dense stand. From above one sees fields and stands of pines; both cultivated and going to waste. So many edges, such small islands of habitat, such an incessant leaching of life through the cuts and peripheries. Despite its pre-owned quality, though, I love my home. That's my mantra. It seems like an ample realm of possibility and obligation, a destiny on an appropriate scale.

This place is where I've spent what will have been, by the time I finish this writing, a very turbulent decade. I've made love on my land, I've dug in it, spit, wept, and bled on it. I've made rituals here. I've planted trees and gardened and been an outraged citizen here, battling against the encroachments of "progress," such as the rapidly expanding landfill one mile due south. For all that, I am only beginning to know these acres. I had lived here for two or three years, walked all over the place before I really saw the venerable beech. It's an incredible old tree at a far corner of the property. It looks like a herd of charging elephants. It takes three of my armspans to encircle the trunk, and its branches, in summer, cast a pool of shadows twenty yards across. When I took Bea, a Chicago bioregionalist and yogini, back there a few summers ago, she said that in India, a tree like this would have a little pile of stones, an accumulation of offerings, at its base. Mine has a stain of New Mexican blue cornmeal in a cleft, rather unbioregional, but a best gesture for now. Aside from the beech tree, this is not the most attractive part of the county to be living in. There's prettier land not far away. It's not that I'm taking my stand here out of pure altruism. My being here also has to do with inertia.

A few winters ago, after I became single, I found myself in a bioregional circle of friends. We were discussing our relationship to place—to this area of Michigan as our homeland—and to our property, those of us who were landowners. I'd come here for ideological and romantic reasons, having fallen in love with a local man, a brilliant bioregional organizer. Visiting him at the height of summer, I fell in love, too, with the Leelanau Peninsula. When the personal love faded, and I was left to my own devices, stuck out in a Scotch pine woods, the infatuation with the land also dimmed. Like most every place of late, this Leelanau Peninsula has been under steady attack by the forces of what is misnamed "development," and, indeed, some of its pastoral charm and remaining natural beauty have been destroyed. But beyond that visible change, I experienced also a change of heart, and a change of regard.

There, in the group, I said that the change felt like marriage when the intoxication fades and the irritation sets in; when the flaws become less tolerable. I thought of all the times I'd looked out on the Scotch pines surrounding me with no ability to see them as they are; seeing them only in terms of their ecological status as nursery-bred alien trees: more twisted, more stunted, and certainly less generous than is actually the case. The knowledge that they're exotics rather condemns them in my mind, which may be why I wishfully thought them to be natives—red pines—for the first few years I lived among them.

As this shameless disaffection spilled out, I said "I'm an exotic here, too." It was an admission of being confused. Like the pines, I don't quite see myself as being optimally suited to land citizenship around here, although the simile only goes so far. Scotch pines are a lot hardier and unquestioning about their immigration than I am. "And I don't know what I'm supposed to give way to, but I mean to hold down some soil meanwhile. Nevertheless," I concluded to my friends, "I believe that for a young American such as I, staying put can be a salutary discipline. To paraphrase James Joyce's heroine Molly Bloom, 'I thought as well here as any place . . . and yes I said yes I will Yes.' "

If I can manage to hold on to my land for a few decades I can at the very least leave it to its own devices and give it time to recuperate. At the moment I have no need or wish to realize pecuniary gain through land ownership. (This was manifestly not the case for the previous owners, who variously logged it, farmed potatoes, and established a Christmas tree plantation, all for some small gain.) Even with such modest ambition, I wonder how I am going to do what I have set out to do, which is to learn and tend my own land, all thirty-five acres of it.

Once my good intentions depended on having a capable male partner to execute them with, someone who could teach me, involve me in healthy, hearty work on the land. It's the outmoded female fate of achieving things vicariously, with the permission or intercession of a husband. Thus I arrived in this place, beautiful and tough as it is, from a half-conscious compulsion to marry. I wanted to annex my husband's talents and abilities. I hate to confess having been so retrograde, but my past is reality, after all. Perhaps it's not necessary to condemn it in order to learn from it. Like Nature, I suppose, it can be killed and dissected as one path to knowledge, or observed and sung as another.

In 1984, then, I took up residence with my ex-husband-to-be in the farmhouse that used to be the swingingest bachelor hovel in these parts. Township 28 North, Range 13 West, Section 26 of Leelanau County is where I wound up dwelling, with him and another now-former bachelor. After a few years of group living, my husband and I built our bright little conjugal house next door to the bachelor estate on a skinny five acres that we purchased from our former roommate. The marriage ended a couple of years after that. One of its more traumatic episodes was our suffering a head-on automobile collision whose consequences upon my body—a badly broken right leg—provided me with a little too much opportunity to observe the organism's long, slow healing process. It didn't help my homesteading much, either.

You divorced readers may well understand that the parties in a divorce are usually insane for quite a little while before, during, and after, a few bubbles off-plumb. The humane and sensible thing would be to have one's affairs put in trusteeship and have oneself committed to an asylum for a couple of years, while the worst of it blows over. Unfortunately, you wind up making all kinds of major life decisions right then. I was fanatically certain about my choices as I spun in that personal chaos; lately I am less so. One choice I made was not to beat it back to California, whence I had come, but to remain in Michigan. During four years of wedlock I had come to like this place and my friends here pretty well, so when the cyclone of divorce cut loose, the eye, for me, seemed to be the house we'd built, and where I stayed. The year following the divorce my neighbor gave me the first option to buy the thirty acres surrounding my place and extending to the south, a chance to become land-poor. The attachment that ultimately decided the land purchase was to the beech tree. In offering me the first opportunity to buy the acreage containing it and surrounding my home, my neighbor mentioned that he had observed that I felt about this tree like Dian Fossey did about her gorillas, and figured I would be none too happy with the thought of someone else possessing it. So I bought the land. I may not have made the right decision, but made it is, deepening my stake in remaining here and in developing some local knowledge.

In late February there are suspicions and portents of spring—odd deluges of heavy damp snow, slight thaws, plunges back to freezing, and slicks of ice. There is an opal undertone in the birch-bark, bronze in the aspens, gold in the great willows away in the wet-lands, and garnet glowing in the fire cherries. I see eagerness in all this splendor, though it may just be that after the long Zen brush painting that is winter, my eyes are so ravenous for color that they are seeing hues that have been here right along. Intimations notwithstanding, there was still a serious mantle of snow on the ground. Lacing up my heaviest boots, I bestirred myself to go out back, imagining that I could slog around through the snow upon occult terrain and take a reading on the world of Scotch pines.

I had lately read, in a collection of his essays, Aldo Leopold's observations on the intensively managed German forests that he visited in 1935. In them he remarked "an unnatural simplicity and monotony of the vegetation of the forest floor, which is still further aggravated by the too-dense shadow cast by the artificially crowded trees, and by the soil-sickness . . . arising from conifers.

"The forest landscape is deprived of a certain exuberance which arises from a rich variety of plants fighting with each other for a place in the sun," wrote Leopold. "It is almost as if the geological clock had been set back to those dim ages when there were only pines and ferns. I never realized before that the melodies of nature are music only when played against the undertones of evolutionary history. In the German forest . . . one now hears only a dismal fugue out of the timeless reaches of the carboniferous."[1]

This assessment was much on my mind as I crunched around amid my welter of Scotch pines. Though I was grateful to be sheltered in a woods, and to indulge the illusion of rugged outdoorsiness, it was more like being a ladybug toiling around in the blades of a suburbanite's zealously cropped lawn. My land has relief and dense, unexceptional vegetation. Long on one species, short on diversity, it's a bastard community.

At the 1992 Society for Ecological Restoration conference in Waterloo, Ontario, I met a landscape architect working for Her Majesty's Forestry Commission in Edinburgh. He has a plan for reinstating native woodlands in certain montane reaches of the United Kingdom which are now conifer plantations. He quoted Wordsworth's judgment upon these monocultures: "abominable vegetable manufactories," and mentioned that some of them consist of Douglas fir, which is a tree of the North American Rockies and the Pacific Northwest. In some places, the restoration will entail planting Scotch pine.

Meanwhile I survey the twenty of my acres thick with Scotch pine and contemplate restoring them, over time, to something more nearly resembling native woodlands. In light of these transatlantic foresting follies and other such snafus too numerous to count I deem that the epitaph of our civilization should read: "It seemed like a good idea at the time."

To fathom the reasons for the presence of Scotch pines in northern Lower Michigan, a region with its own great native pines—red, white, and jack—and of the gross transformations of our woodlands generally, I consulted a couple of books that had found their way into my possession: a U.S. Department of Agriculture yearbook of agriculture on *Trees*, circa 1949, and *An Introduction to American Forestry*, circa 1960. This latter was fairly appalling in its utilitarianism. The only good forest, it would seem, is a managed forest. "Wild" fires are wasteful. Both books, the U.S.D.A. anthology particularly, were imbued with the optimism that technology would always redress Nature's ignorance. This continues the optimism of the EuroAmerican plant explorers and transporters of the eighteenth and nineteenth centuries. For these naturalists, moving tree species across oceans to experiment with them in new settings seemed a most worthwhile way to boost the Earth's productivity for human purposes. Exotic trees were miracle crops from abroad, and there was no introduction that couldn't be successful.

Scotch pine, *Pinus sylvestris*, is native to a broad swath of northern Europe, from the eponymous Hebrides to Russia and Finland. This tree is a trouper—hardy, prolific, a ready colonist of poor soils, albeit something of a perpetuator of soil poverty. It's tough and can withstand the assaults of industrial air pollution. Where it belongs, Scotch pine historically has attained heights of 70 to 120 feet and girths of from two to four feet. The inner bark of Scotch pine is edible (as the porcupine out back has demonstrated). In Kamchatka, which lies between the Bering Sea and the Sea of Okhotsk, this pine phloem constituted part of the human diet. In the British Isles, Scotch pine timber was favored for small, nimble yachts, for masts and spars. They say it makes a handsome forest, not altogether inhospitable to other trees such as birch, aspen, and, occasionally, oak.

The tallest, healthiest Scotch pines I've seen in these parts are at most thirty feet high and their contorted upper branches exhibit a papery, saffron bark. What they are doing here, mainly, is being farmed for Christmas trees and sometimes planted as windbreaks. Tended properly, Scotch pines do make pretty Christmas trees. They are potentially shapelier and more fulsome in their foliage than the native jack pine. Red or jack pine trees, however, would be the native talent likeliest to play the colonizing-of-the-hopeless-dry-sandy-soils role if that were the true objective of such afforestation. Jack establishes itself well but is too quirky and gnarled a character to function as a plantation Christmas tree. It's darkly, craggily handsome, and its cones require periodic burning to release their seeds.

Hence the Scotch pines that surround me. Although their aerodynamic seeds can travel great distances (one researcher found seeds in Alpine snowfields ten to fifteen kilometers from the nearest trees) to enlarge their territory, I haven't noticed pines germinating more than twenty feet beyond the borders of my stands. Within the stands there's enthusiastic reproduction—wherever there's room, one encounters small Scotch pines of a few years of age. And the plenitude of the trees' seed production is attested by the thousands of pine seedlings that make an ankle-high, spring-green haze in the bark litter that covers the ground around my house.

Apparently my pines pose no imminent danger of swamping an adjacent native ecosystem. Such a stable climax community should, in theory, be impervious to such invasions, but it's all theory because there isn't a climax community on my property. These Scotch pines so densely occupy my acreage, however, that they are perhaps retarding its succession back to hardwoods by their thick shade, by raining down a slow-to-decompose, soil-acidifying mat of needles, and by their hungry, pervasive network of roots. Wherever I've plunged a spade into my earth, I've hit an inch-thick pine root. In a regime of benign neglect, the intruder pines, unthinned and untended, eventually would shade themselves to death, and would be succeeded by hardwoods. For a dabbler in ecological restoration, though, eventually isn't soon enough. I want to tinker with the successional process, to nudge it along by selective thinning and planting.

Excepting the fact that they don't belong here, there is much to love about these pines. The very terms *anemochoria*, which means wind-dispersion, their modus operandi of seed-scattering; and *anemophily*, which means wind-pollination, invest their enterprise with a certain charm. In the springtime, when the Scotch pines are flowering, all the surfaces in my house acquire a fine sulfurous coat of pine pollen, deposition of the tiniest fraction of the tons of airborne pollen grains the male flowers collectively produce. Verily a great spawn!²

These Scotch pines, *non sequitur* though they may be to northern Michigan, indeed to North America, are sheltering me, and I've grown fond of them. That realization stole upon me unawares one moonlit night as I pulled in to my driveway and to the welcoming presence of my house. The pines rise up on all sides, protecting me from view. If a weed is, as the botanists say, both "a plant out of place" and, as Emerson had it, "a plant whose virtues have not yet been discovered,"³ then Scotch pine, on my land, is one but not the other. It is far better than nothing for vegetation and its crowded misrule has begun to please me. Still, I would like to foresee a time—beyond my lifetime, probably—when the deciduous idea reasserts itself here. My pines are so thickly planted and ill-maintained that it is possible to enter them and get lost, to enjoy the illusion of being the only person in the world, safe and hidden. It's underbrush, not forest, not a grove, but the *maquis*. Prowling it I'm not a druid but a denizen, crouching and twisting to make my way through the branches and deadfalls, breasting dead twigs and cutting an erratic trail through the farmed-out potato fields of yore. Such are the pleasures that sickly land affords. I am glad that Thoreau did not live to see weedy landscapes like this, and didn't have to make shift with the deserts we have created to find himself a world.

I went for a barefoot walk one summer evening. One of the great straightforward pleasures of this season is less-fetteredness, the clemency of temperature making it possible to expose one's skin to the world, which world seems at these times to be thick with blood-sucking insects also rejoicing in the exposure of arms, legs, and fore-heads—pale pink fields of plenty. A recent spate of rain greened up the mosses here and there, and part of what lured me out was the lux-ury of treading on them for a few paces before crunching across the next patch of lichen. The sheep sorrel was flourishing and succulent, a balm to the soles.

I think I know what this old farmed land should be able to be again. It once knew how to produce a hardwood forest. Low places in the fields where the plowed topsoil washed down produce dense grass, even make sod. Right now the cleared land, which constitutes the majority of my neighbor's estate, and a little of mine, hosts some tough customers, a mix of weeds and bunch grasses: spotted knap-weed most conspicuously, vetch, oxeye daisies, pepperweed, and Saint-John's-wort, or *Hypericum*. A collection of plants so tough you see their skeletons throughout the winter. There's an occasional black-eyed Susan or Deptford pink; we see hawkweed and cinquefoil, horse-mint and rosettes of mullein; a patch of blackberry canes cloning out-ward. Closer to the woods, there are bracken ferns under the cherry trees, and a few grasses. This evening I saw a lichen that looked like daubs of cement strewn by a careless mason, primordial cover of rather grim aspect. Complexes of plants assorting themselves in rude communities according to the haphazard conditions resulting from clearing and farming and abandonment, to say nothing of a few years of strangely dry weather.

One early spring afternoon's walk, as I zigged and zagged down the alleys between the rows of pines, tacking to avoid the more impossible tangles of limbs and fallen trees, I emerged into a clearing close to the southern edge of the pine plantation, and within sight of the top branches of the great old beech. Around the perimeter of this open space there were scores of beech seedlings, nestled at the bases of the scanty young pines, or making partnership with pine seedlings, their coppery dentate beech leaves trembling and clattering in slight stirrings of wind. That these babies might be the offspring of the noble old tree and will be carrying on her demonstrably strong lineage is a happy thought. These descendants are doing just what Rick Moore, the Soil Conservation District Forester, predicted during his survey of my land: recolonizing open places. They're outriders of a new old order and harbingers of an eventual return of richer, deeper soil. After rejoicing in the presence of all these youngsters, I walked on to visit their *magna mater* and perceived for the first time that the land rolls away gently on all sides from the low hill she surmounts. This old beech seems to hold the high ground, at least on this square mile of the world.

When I am prowling about back on my land, I enjoy a strange sense of protection. The landscape is a contrary, brittle maze in which the botanic surprises are few. Even so, those Scotch pines are sheltering other life. The beech babies in their shade are fine, as are the sugar maples.

Forging a relationship to this terrain, to my preserve, is a long, slow story. This is my habitat, a place where I can enjoy my animal nature, where I can await the revelation of a mystery, where I can puzzle out the happenstances—an edging of goldenrods around a bare place, fresh-kicked earth at a burrow, or the annoyed natterings of the chickadees, one talking with its mouth full, reacting as my cats and I cross a clearing and fold ourselves back into the little birds' pines.

I'm trying to purge my interior monologue of the prefix "It's just a" as in "It's just a blue jay." Is it fair to decree second-class citizenship for the species that are able to adapt to human disturbance and intrusion? Disdain for survival skills ill becomes a nature buff, and the blue jays, with their glorious plumage and wide vocabulary, certainly enliven the scene.

Taxonomy, I think to myself, is an impediment to perception. Of course it's important to know which group an organism belongs to, and the differences among families, genera, and species are wonderful in themselves, lavish testimony to the theme of variation. But for a person already wary of equating experience with descriptive lines of type, this disinclination to do the work of naming things, while at the same time regarding them as somehow less real without their official labels, has turned my immediate surroundings into a stack of unfinished homework late on a Sunday night. "What is that?" and "It's just a" are the small phrases standing between me and epiphany, or simple observation, never mind the big religious experience.

Yet in this as in other tasks involving language, I respect precision and the richness accurate language affords. Specific identifications of trees and birds and shrubs and mosses give a definite sense of the living, teeming world. It may be that we are in the habit of destroying biodiversity because industrial civilization runs on commodities. If trees are just cordwood, not sugar maple, silver maple, red maple, and box elder, our sense of the world's texture is reduced, and we can't come up with a detailed answer to the question, "What have we got to lose?" Does making an identification enhance or distract from the possibility of animistic recognition? The problem of knowing confronts me now, a mid-life epistemology crisis.

Keying out a plant is not my idea of a good time, but it's a necessary chore for the aspiring reinhabitor. One can't help but appreciate, finally, the methodical, invaluable ordering imposed on botany by systematists. How cleverly they lead you onward to the prize of all but certain knowledge!

In front of my studio there grows a little shrub whose black berries have on autumn occasions drawn a jovial flock of cedar waxwings to feast. Naming the shrub provided an occasion to utilize Gleason's *The Plants of Michigan* thus: Start with "Shrubs or woody vines without true tree habit, or attaining heights of less than 20 ft." Go to specimen with leaves and flowers. What kind of leaves? Broad flat leaves. Not twining or climbing. Arrangement of leaves? Alternate. What kind of leaves? Simple. Normal in shape, green in color, deciduous. Flowers? Not in catkins with petal-like parts. Flowers white with conspicuous petals. What kind of petals? Separate. How many? Five. Corolla? Regular. Stamens? Ten or more, twenty-four in Rosaceae. Ovary? One, permanently enclosed within receptacle. Habitat? Native species shrubs growing in woods, fields, or thickets. Shrubs without thorns, and, the clincher, "leaves tomentose [covered with dense, woolly matted hairs] beneath," makes it *Pyrus arbutifolia*, the chokeberry.[4]

The identification process depends on bifurcation and elimination, arriving at the name by tight focus on unambiguous details. With discussion of context and habit left to other kinds of works, keying out a plant is, in many ways, a mere beginning point of knowing.

Grass flowers are inconspicuous, to say the least. But there is something positively orgiastic about their mating season, these grasses of which all flesh is made. One day, at a time when many of them were blossoming, I decided to try to identify the species in my territory with the help of Lauren Brown's handy *Grasses: An Identification Guide*. Northern Michigan is meant for forest, not grassland. The fire-dependent prairie provinces, whose plant communities engage the attention of burgeoning numbers of restorationists, begin a few hundred miles to the south. But old fields, of which there are many hereabouts, and pastures, places cleared of their forest, host mixtures of grasses from around the world.

Unsurprisingly, then, the majority of the grasses I found around my place were exotic to North America: smooth brome (*Bromus inermis*), Canada bluegrass (*Poa compressa*), orchard grass (*Dactylis glomerata*), meadow fescue (*Festuca elatior*), English rye grass (*Lolium perenne*), and crab grass (*Digitaria sanguinalis*). At least the others would be found on this continent: quack grass (*Agropyron repens*), purple love grass (*Eragrostis spectabilis*), red top (*Agrostis alba*), green foxtail (*Setaria viridis*), and old witch grass (*Panicum capillare*). Timothy (*Phleum pratense*) may possibly be a North American native that was sent to Europe shortly after its discovery and then returned.[5] This provisional list tells me that Michigan farm country, part of the "New World," is also what Alfred Crosby in his *Ecological Imperialism* terms a "Neo-Europe."[6] The grasses and herbs of the old country are jostling here with the locals in a landscape structured first by axes and oxen, then by plows and cows; and by plants with long experience in the Old World of learning to cope with these.

I woke at six and decided to go out walking during the sunrise, hoping to hear the songs of birds returned for spring, and mating. The whole season comes as a surprise. It's as though the adaptation to winter entailed some amnesia as well. By December I'd forgotten that this magical return of green to the dull and trammeled earth was in store. Now the young beeches will let go of their leaves, which have been clattering lightly in the winter's many winds.

Incidents of an early April morning: the dim flash of the disappearance of a white-tailed deer who noticed us (for the cat Tyrone was with me) before we noticed her in the morning's dark. We travel in a cone of silence. Only the boldest singers venture into it. It was cold enough that I wanted to keep moving, so I didn't stand still long enough to allow the beings around us to regain their trust and act themselves again.

Down by the pond, around the base of a poplar whose top may have been snapped by a wind, I noticed a scatter of lurid white chips, quarter-inch bits of pithy heartwood drilled out by a woodpecker. Downy and Hairy are the species I see most frequently. I tried to imagine what it would be like to hammer away at a snag with my jawbone in order to establish a nest. A brain-rattling proposition, but the dapper little woodpeckers are made for it.

On the day's ramble, I also learned that you can hear ants. There are two or three big anthills out back, perhaps a foot and a half high and two and a half feet in diameter. Somnolent a few weeks earlier, their surfaces were now teeming with big, two-toned ants, ants coming and going through dozens of entrances, single workers and two-ant teams hauling fir needles, scales from pine cones, bits of things edible down into the hill. The mounded surface of this Jericho was now in nervous motion. There was such a number of these tiny beings that you could hear the cumulative noise of their activity, a small stirring audible over the wind and bird song and the ever-present growling of engines in the distance.

Except for some shady, wind-sheltered swatches in the pines the snow was gone. Out in the sunnier scraps of hardwoods, though, most of the beech leaves driven down by the weight of snows and compressed into a crackling layer of papyrus are now exposed. Through this the wiry, aggressive beech shoots are piercing their million ways to the light, as are tips of trout lily leaves. For every beech seedling that succeeds unto treehood, about two thousand fail.[7] Quite an unsentimental ratio.

W inter snows make it possible to see animal tracks nearly everywhere my eye alights, and in great variety. Taken altogether, it can look like a rush hour, but each passage is inscribed in a separate moment. It is gratifying to see all these signs of life, and qualifies my drear sense of the plant life of my land. These Scotch pines may be aliens, but they are cover, and the two- and four-leggeds appreciate their service. Mammals, who account for the most conspicuous tracks, are obligingly large enough to identify. Insofar as I've kept a naturalist's notebook these last few years, I've been describing meetings with fellow mammals, and reading up a little on their biology.

Sylvilagus floridanus

One late February day I flushed a grouse and frightened a cottontail, who bounded quickly out of view. On my return to the studio, I consulted the books. Rollin H. Baker's *Michigan Mammals* calls "saltatorial locomotion" (the scientific term for "hopping," I guess) a characteristic of the family Leporida, order Lagomorpha, to which the cottontail rabbit belongs. Vegetarians, rabbits do a lot of chewing, hence the cottontails' chisel-shaped incisors continue growing throughout their short lives. With skin too thin to be of value in the fur trade, cottontail rabbits went largely untrapped, eventually to serve as the "farmboy's big game." Reading further, I am a little disgusted to learn that rabbits re-ingest their fecal pellets, but it's all in the name of economy and complete utilization of food.

The speculation is that the cutting of southern Michigan's hardwood forests helped create more of the edgy, brushy disturbed habitat that cottontails fancy, and that in their expansion northward into this newly favorable terrain, they learned to utilize the burrows of other mammals, like the woodchuck, in order to survive the colder winters. Three rabbits per acre is an abundance; adult cottontails tend to give each other space. Late winter is the time of year when their mating season begins. Females may have as many as four litters of four to seven young in a season, a reproductive rate reflecting the fact that life is chancy for the cottontail. Three years is a long time for a cottontail rabbit to live. Small wonder, considering their legions of enemies: long-tailed weasels, feral house cats, domestic dogs, red foxes, red-tailed hawks, northern harriers, barred owls, great horned owls, crows, black snakes, gray wolves, and snowy owls. The automobile, not a predator in the true sense, is a major cause of cottontail mortality.[8]

Deep-layered and crusted snows of the sort that lay about in February and sometimes March serve to lift cottontails closer to their suppers, the barks, twigs, and buds of various woody plants. Nervous-seeming and elusive, the cottontails must find things to like about the Scotch pinery—it affords them good cover and ready access to my garden, which, despite its feebleness, presents them in due season with an array of gourmet treats and a standing invitation to nip things off neatly at the top of the stalks, as is the rabbit's wont.

Erethizon dorsatum

I went for a late winter walk in my woods with Michael Jarvis, who, having studied the art of tracking, was to share some of that skill in observation with me, thus helping me to develop a description of what is here. It was not such a cold day, with the temperature perhaps in the twenties. There was a foot or so of snow on the ground, with much deeper drifts in some places. The sky was gravid with still more snow. The sun's muted light graded between gold and lavender.

Before we headed into the woods, I showed Michael the aerial photograph of the land. On the ground, especially in the thick of the Scotch pines, the rectilinearity of these plantations and boundaries is not so obvious as it is from the air. As we were about to plunge east through the phalanx of Christmas-tree spruce that borders a portion of the pines, we noticed that the male cat Tyrone had come along for the expedition. It was snowing almost imperceptibly, just enough to add critical mass and cause random clots of snow on pine branches to drop suddenly, trailing ice crystals that would hover, like a ghost of the event.

We crossed my parcel to the red pine plantation to the east. The perfect columnar avenues of trunks seemed cathedral-like and partly unreal. Past those we entered a little patch of hardwoods featuring another old beech, this one bearing an inscription dated 1933. Most often when we looked on the younger beeches and pines we saw patches where bark had been stripped and trees girdled, strongly suggesting the activity of a porcupine. Michael showed me the claw marks on the slaty-smooth bark of a beech tree, faint crescent abrasions indicating the porcupine's climb and descent. We sat quietly for a while on a ground cloth in the snow, with the cat's purring in my lap supplying a soft foreground rumble. The whinings and grindings of the heavy equipment at the landfill made a noisy backdrop to the middle ground of subtle sounds surrounding us, the sounds of branches rebounding, released from a weight of snow, of parchment beech leaves rustling. In the intervals between waves of sound it was still enough to hear the gulp when one or the other of us swallowed.

On this walk Michael taught me how to look for animal hairs in the vicinity of their trails. Even without the Sherlock Holmesian fun of trying to deduce what the critter was from such minute and obscure clues, the practice of stopping and carefully examining a particular spot outdoors brings a tremendous amount of detail into view. Standing still and looking restores some extent to the world, even as traveling in a vehicle or just striding purposefully diminishes its texture.

In embarking on this project of learning about the community on my own land, and in watching the psychology of my relationship to it, I had imagined that I could, presto change-o, turn myself into a naturalist and do a biological survey of my thirty-five acres. I would be like a good detective collecting names and whereabouts, especially names and positive identifications. For a start, I went to the library and amassed shoals of books—books on mammals, books on forests, books on Indians, books on geology, and soil maps. The wildlife book began by telling me about the distinctive features of the porcupine's skull.

Back in the woods Michael had surmised that we might see a porcupine somewhere up in the neighborhood of all those munchings, and so we did, about twenty feet up in a pine, a torpid, luxuriant black and ocher wad. We looked at it through binoculars, got a closer view of the depth of its pelage and the abundance of its quills. There was just no telling whether its skull conformed to type.

Vulpes fulva

Making my way south along the deer run behind the garden in a light September rain among the scruffy pines, I passed half a dozen kinds of fungus. I veered southeast toward a patch of dying pines whose lower branches were grizzled with silver lichen, noticed the hammered gold of a young maple, coppiced by herbivores perhaps. I noticed a little patch of goldenrods going to seed and in this wild flower bed saw a fox skeleton. The skull was about as long as my pencil. It had huge elongated orbits. The lower jaw lay in two pieces. There was what looked like a collarbone, a rib or two, a shoulder blade, some greenish scraps of fur reposing among tumid sorrel leaves. As I stood making notes, blue jays gathered and began to chide.

The next day while resting I touched my own jaw and was reminded of that fox's lower mandible lying in two saw-toothed parts, and of those haunting sockets of bone emptily confronting me. Was this the red fox that came barking at my door one morning the previous April? Its fur was an off-color yellow, not the rich rufous pelt one associates with red foxes, and it had some black spots. Foxes around here are afflicted with mange. And a chicken-keeping neighbor is said to set out poison for them. On two consecutive days this fox had strayed near my house, heralded by a crow the first day, announcing itself with hoarse yaps the next. At the time I fancied I was receiving a totem harbinger. Now it appeared more likely that the fox had been too sick to steer clear of my dwelling.

Five months later I stood looking at the few scraps of fur green with decay and wondered whether that unusual patch of goldenrod had sprung up as a floral tribute nourished by the slow dissolve of the fox's body, or if the fox had sought out a bower in which to die.

Tamiasciurus hudsonicus loquax

Tyrone has killed many red squirrels, bringing them into the house to devour them on the bathmat, daintily severing the paws and tail, leaving a portion of the gut untouched. The bird feeder is the squirrel's downfall, as well as the occasional chickadee's. It isn't fair to lure these creatures close with a bonanza of food, making them vulnerable to the depredations of a supremely keen solitary killer. Indeed, belling the cat might be one of the most ecologically restorative things I could do.

One day Tyrone brought in a red squirrel that wasn't quite dead. While I scolded the cat, the squirrel erupted from its fangs and shot up the brick face of the chimney. I donned my work gloves and went upstairs where I could reach out from the balcony and possibly grab the creature. When I got near, the squirrel leapt to the floor, plummeting ten feet onto a hard surface. Its impact left a bloody splash. I chased the squirrel into the bathroom and finally out the window to freedom. The little squirrel's energy was enormous. It may only have weighed eight ounces, but its fierce flight filled my whole house with unpredictability, fear, and a will to escape. It was as much an encounter with a wild animal as if a bobcat had wandered through.

These feisty little squirrels mate in late winter and at the height of summer. Four to seven young are born at a time, and the mother stays with them until they're nearly grown. They eat nuts, pine cones, mushrooms, bird eggs, meat, sap, and freshly deposited kitchen scraps from the compost heap.⁹ Sometimes, as I look out my second-story bedroom window, it seems that every Scotch pine branch is bouncing from a red squirrel's passage, and looking out the kitchen window to the west I often see these aerialists doing the *salto mortale* across a yard-wide gap between two treetops. It's better than Ringling Brothers, Barnum and Bailey.

Canis latrans

One morning a coyote appeared grinning behind the house. With a lush tawny pelt, lanky limbs, bright eyes, and what I took for a smile, it was there and it was gone in a split second. That apparition got me thinking about the coyote's latter-day eminence on the North American continent, how coyotes are extending their range; not going away, but going farther, and are the big feral canine presence in lands no longer visited by the dire wolf. "Today," says Rollin Baker, "the coyote is the dominant terrestrial carnivore on most of the continent, with the human as its greatest enemy."[10]

Didelphis virginiana et alii

My neighbors came over for brunch one early summer Sunday, and we got to talking about the local mammals. They mentioned that an opossum had been visiting their compost heap. The ability of this marsupial to mingle with human settlement, indeed to thrive in disturbed lands, apparently accounts for the northward extension of its range, although the migration has its costs. The books say that possum ears and possum tails are often damaged by frostbite. The possum refuge of choice is a den dug by a striped skunk or groundhog. Like contemporary Michigan homebuilders, opossums value insulation as a means of staying warm through the long winters. Half a dozen bushels of leaves may be gathered and stowed to make a possum home cozy.[11]

Also noted: a striped skunk, and a bat, whose species I didn't identify (it was dead; a specimen acquired, to my chagrin, by Tyrone the terrible taxonomist). Mouse tracks in the snow, looking like curving lines of stitches between grass stalks (white-footed mouse would be my guess, but I've not yet seen one on the move). Short-tailed shrews (also courtesy of the lethal cat), with their minutely glittering obsidian eyes. A mole, deceased for no apparent reason, its cool dead body wrapped in infinitely soft fur, its fleshy pink flipperlike forepaws calling up an image of the creature blindly pushing aside clods of soil in its subterranean foraging. Raccoons also of course. These bold unpleasant omnivores scuffle around the house in spring and fall, manifesting the nerviness that's kept them plentiful.

White-tailed deer are everywhere, yet manage to be nowhere when hunting season rolls around. Once I saw impressions of their sleeping bodies warmed into the snow—lush hollows left by shoulders, slender flutes melted by forelegs. Deer, cottontail rabbits, and porcupines are all implicated in the scarcity and pollarding of the hardwoods that are attempting their comeback amongst the Scotch pines. As winter wears on, hunger drives these herbivores to gnaw and crop any woody plants that are palatable in the least. Porcupines, with their taste for inner bark, will girdle trees, and are the primary suspects in a sugar maple girdling incident out by the compost heap.

They're all garden-variety mammals, the biggest vertebrates in this depauperate fauna, streetwise legatees of wildness. They are my closest companions—hot-blooded, viviparous—on this disturbed ground.

One mid-July afternoon I went walking to find the spirits. The previous day's weather had affected me terribly—it was hot and humid, tumultuous and windy, with a hint of violence in all the harrying of leaves and branches. I was immobilized, prostrated, both dreading and praying for the forecast severe thunderstorms to begin, to precipitate some change, any change, in the barometric pressure and charged, louring skies. I was ready for a tornado to blow up my studio and for lightning to blast my house, anything to get it over with. I spent the long daylight hours either pacing or draping myself over various pieces of furniture, trying vainly to keep my head erect enough to read, if not write.

Last night's hissing curtains of rain and just-distant-enough thunderclaps and lightning flashes slaked, refreshed, and cooled the world. I slept under an open window and was christened by a mist of droplets kicked up off the sill. Morning came and the cool remained, as did a degree of the earlier lassitude. There was fresh washed air admitting pure sunlight, and breezes again, these winds not seeming so aggressive, but tossing foliage and currying grasses, a playful animation of this green world. It became an extraordinarily beautiful day. Thanks to whatever constellation of forces—volcanoes, El Niño, the jet stream, the Gulf Stream—what looked like the inexorable march of global warming had paused for a moment.

As I walked out in the mitigated heat of the afternoon, I was as usual noticing the interplay of degradation and regeneration on my land under its current regime of therapeutic nihilism. (It's the school of land management which holds that we are too ignorant to know what is the right thing to do with an ecosystem. Therefore it is best to follow the Hippocratic caution on First Do No Harm.) The old field part of the acreage continues to succumb to the onslaught of spotted knapweed, which is insinuating itself in places where a weed, by rights, should not. As in among the thick pelt of grasses at the lower points in the field. The knapweed is doing nicely, thank you, and is not the only happy alien. Sorrel went crazy this year, and oxeye daisies—galaxies of them. Bladder campion abounding. Clumps of purple vetch. *Hypericum* has burst into bloom, Goat's-beard has gone to seed, its windborne parasols cruising lazily about. Not all come to earth. I watched one drift, get snagged by a pine bough. This year I noticed an abundance of Deptford pinks again. It seems they were missing for a while there.

Walking back to what used to be a love-grass meadow, I was disgusted to note that it too was thick with knapweed. Only a few years ago, I was walking back there with Erin and Annie, the neighbor girls, just when the grasses' tumbly seed heads had ripened and begun breaking off and dancing about, catching the low-angled sunlight and making a nimbus on the clearing. It was soft and inviting and Annie got the fine idea that we should "bury" Erin, cover her with grass panicles. Which we did, with much tickling and giggling, then each took turns being buried ourselves. The effect was a little like Snow White's glass casket—ethereal and not-unpleasantly eery.

—Well that'll never happen again. Knapweed is stiff and most unpleasant, raw and angular, a little bit prickly. That little clearing, for now, is waste.

I hope it is not conceit to attribute my ongoing residence here to Thoreauvian impulse. There was an honest urge toward voluntary simplicity, toward freedom, and toward the opportunity to wander idly in a little woods under my protection and care, that bound me to this land in northern Michigan. I was prompted also by animism, by the sense that it is all alive, that this place and I have a relationship to work out, that there are certain realities we must confront together.

Thoreau may have been among the first Americans to notice that what we would now call ecological restoration was called for.

"We are a young people," he wrote in his *Journal*, "and have not learned by experience the consequence of cutting off the forest. One day they will be planted, methinks, and nature reinstated to some extent."[12]

Part Two

Chapter 5

The Leopolds' Shack

During the Depression, Aldo Leopold, a University of Wisconsin professor of game management, obtained a blown-out "Sand County" farm about an hour by car distant from Madison, where he taught. What might have seemed to the disinterested observer to be just a hunting camp in the country for a professor and his family proved to be the alchemist's retort, the place where an encounter with degraded land crystallized Leopold's thinking on the relationship of people to land and eventually led to his creation of an American classic—*A Sand County Almanac*. The book's spirited proposal of an ethic and "science of land health" has, since its posthumous publication in 1949, been a touchstone of ecological awareness in the United States.

The vessel for the thinking-through of *A Sand County Almanac*, and the setting for many of its essays, is the Shack, a remodeled chicken coop set amidst acres of ill-used farmland bordered by the Wisconsin River. At once celebratory and elegiac, the essays and sketches in *A Sand County Almanac* assessed the state of the land community in mid-twentieth-century America and found it diminished, but never lacking in engrossing details. Leopold reflected on the failings thus far of the morality of the human relationship to land, and proposed a simple, if radical, corrective in the land ethic: "A thing is right when it tends to preserve the integrity, stability, and beauty of the biotic community. It is wrong when it tends otherwise." From 1935 until his death in 1948, Aldo Leopold spent weekends at the Shack with his family, happily experimenting in ecological restoration. Each spring the Leopolds planted thousands of pines and carried on a variety of other projects—prairie restoration, tamarack transplantation; they even propagated sumac, which in many waste places volunteers as a woody colonizer. From being a dustbowl-era ruin, so barren that you could "see a mile in any direction" (in the words of Leopold's eldest daughter, Nina Leopold Bradley), to becoming again a verdant tapestry of riparian woodlands, sloughs, oak and pine woods, prairies, savannas, and sedge marshes, the Leopold lands have enjoyed a second chance. The renaissance of this place—as significant to the American heritage as Thoreau's Walden Pond, but, mercifully, far better protected—is a tale of applied hope and intelligence—and of the land's forgiveness. The ongoing lesson of the Sand County farm is that anyone having any plot of land to care for—backyard, back forty, community park—can initiate a similar process.

A Sand County Almanac is suffused with affection for distinct beings—a chickadee, an oak struck by lightning, a wolf killed forty years earlier by a much callower Leopold, a tiny plant called *Draba*, a woodcock peenting at dusk, a relict silphium sending up flower after flower until it must at last give up its lonely ghost of a prairie life. Leopold turned his observations of the happenings on his lands into timeless literature. The study of the periodic phenomena of nature—arrivals, departures, blossomings, matings, hatchings, ripenings, fallings, the identity of the first singer of morning and the last at night—is called phenology, and Aldo Leopold's rigorous field notation of such goings-on was a lifelong pursuit. Yet the scientist Leopold's well-schooled observations are passionate. *A Sand County Almanac* is deeply thought and deeply felt. One reads it and is persuaded in heart and mind that love and honor must find a way back into our relationship with the land and its life.

Fierce though he was in arguing against stupidities, and any further abuse of the land, it seemed never to have occurred to Aldo Leopold that we could not replant the tree of life, if we chose. His hindsight didn't darken his foresight.

As described by his biographers Curt Meine and Susan Flader,[1] Leopold was a person of remarkable energy, sensitivity, and dedication to ecology as both science and supreme good. He was a tireless field biologist and enthusiastic outdoorsman whose tastes in Nature developed in favor of the wild. Perhaps this passion, which his scholarship substantiated, originated in his love of hunting, which continued for much of his life.

By the age of forty-eight when he acquired the Sand County farm, Leopold had done a stint as a forest service supervisor in New Mexico, worked as associate director of the U.S. Forest Products Laboratory in Madison, had spearheaded the movement to establish a national system of wilderness areas, written the classic *Game Management*, and finally secured a chair at the University of Wisconsin at Madison. He had, with his beloved wife, Estella Bergere Leopold, established a family of five children. (All five would become natural scientists and four would be elected to the National Academy of Sciences.) Not long before purchasing the derelict farm, Leopold had become involved with a lively community of biologists in the creation of the University of Wisconsin Arboretum. He would serve as the Arboretum's director of animal research. The Arboretum project, an experiment in reinstating natural plant communities on land that had been used for farming and grazing, was such an engaging pursuit, apparently, that Leopold was prompted to acquire some land of his own where he and his family also could practice what would come to be called ecological restoration.

As a pioneer ecologist, Aldo Leopold peerlessly articulated the reasons why the health of the entire biotic community must become part of ethical concern. "If the biota, in the course of aeons, has built something we like but do not understand, then who but a fool would discard seemingly useless parts? To keep every cog and wheel is the first precaution of intelligent tinkering."[2] Conservation and restoration would be the most obvious and effective way to save the parts. These tasks are twofold: there is the objective reality of the health, or pathology, of the land that must be tended to and the subjective task of forging a new relationship with the land. In the essay "The Land Ethic," Leopold alludes to Odysseus's hanging of the slave girls, and impugns the concept of land as mere property:

> When god-like Odysseus returned from the wars in Troy, he hanged all on one rope a dozen slave girls of his household whom he suspected of misbehavior during his absence.
>
> This hanging involved no question of propriety. The girls were property. The disposal of property was then, as now, a matter of expediency, not of right and wrong.
>
> Concepts of right and wrong were not lacking from Odysseus' Greece: witness the fidelity of his wife through the long years before at last his black-browed galleys clove the wine-dark seas for home. The ethical structure of that day covered wives, but had not yet been extended to human chattels. During the three thousand years which have since elapsed, ethical criteria have been extended to many fields of conduct, with corresponding shrinkages in those judged by expediency only. . . .
>
> There is as yet no ethic dealing with man's relationship to land and to the animals and plants which grow upon it. Land, like Odysseus' slave-girls, is still property. The land relation is still strictly economic, entailing privileges but not obligations. . . .
>
> The land ethic simply enlarges the boundaries of the community to include soils, waters, plants, and animals, or collectively: the land.[3]

Giving moral standing to nonhuman nature—the philosophy now called deep ecology, or biocentrism—remains visionary. Respecting the "personhood" of each member of the biotic community will entail a transformation of our material life, as well as of our beliefs.

Leopold's intellectual zeal sought tangible expression. Throughout his career, he was a skilled and dutiful participant in the making and shaping of natural resource and land management policy at every level of government, from the federal to the county. In his writing and speeches, Leopold anticipated a slew of ever more critical concerns: wilderness defense, ecological restoration, the ultimate necessity for an ethical, rather than merely statutory ground for all conservation, the essential role for citizen-scholars of natural history, and the need for individual responsibility as the foundation for a new relation to land. These basic ideas, formulated by Aldo Leopold, have helped undergird much of the ecology movement in more recent decades; his vivid, sonorous prose has long inspired it.

However attenuated our sense of it has become, all living derives from the land. Our persistence on this planet and any hope of future prosperity depends, as it ever has, on each of us learning to become what Leopold termed a "plain member and citizen" of the land community. If we are to restore the land to health and our economy to a fitting scale, it becomes ever more necessary for us to reacquaint ourselves with the land, and to work with it in love and understanding.

Unlike many environmentalists of my generation, I came late to *A Sand County Almanac*, not reading it until fifteen years after the first Earth Day. Now my yellowing paperback copy is as thoroughly used as a fundamentalist's Good Book. Because Leopold's great work is so germinal a text for land-healing, I obtained permission to visit the Leopold Memorial Reserve, which comprises the Sand County farm, with the Shack. My ostensible purpose was to write about the restoration there, but really it was a pilgrimage. After an introduction to the place, I was given permission to wander on my own for a couple of days. Those days felt like great blessings. The Reserve is a powerful literary-ecological shrine, but it is also simply a beautiful, interesting, kindly extent of land.

I went for long walks along mowed trails. Strolling in and out of shady places and openings, from wet down by the river to dry upland and back again, I began to note the gradual shifts in vegetation as I went, the changes in the flora carpeting the path—extents of some little mint, then closest to the river a trailing thing with paired oval leaves. Plantain in lots of places. I tried to follow a deer trail and wound up threading my way gingerly through a dense understory thicket consisting mainly of prickly ash, learning how it deserves its name. Old oaks, hickories, and maples loomed overhead, rosy oak seedlings unfurled their new leaves shin-high.

The first thing that befell me on my first day's stroll was the sight of a robust coyote farther down the path. Very healthy and lush-pelted it seemed to be. Walking on a while longer, I paused to sit on a "Leopold bench," a simple settee of Aldo's design made of six two-by-eight driftwood planks salvaged from the river. Enjoying the sounds of life all around, I learned I was in fabulous mosquito country. Their whining, too, is part of the chorus. I heard what was perhaps a wood thrush. Also a crow, and deer snorting in a waist-high thicket. The foliage was stippled yellow with dew-gathered pine pollen. Interspersed through the woods were some hefty white pines, dating from a time when deer were less numerous and therefore less likely to have browsed out the pine seedlings. Off in the distance, the interstate thrummed.

The moment I approached a clearing by the river, a bald eagle took wing, with a hiss of air flight-combed through its feathers. For a second I wondered what this lordly bird could be, then the dazzling white fan of its tail flashed the field mark of our national totem. Later, as I sat at a camp table writing, I was investigated by three different kinds of spiders, including a natty jumping type in black and white with emerald eyes. I waded into the tea-colored Wisconsin River, testing the current, which impressed me with its force. I decided therefore not to attempt to cross it, or even venture into the main channel, hoping this was prudence rather than cowardice. I just figured they didn't need any drowned authors floating past the Shack.

The original extent of the Leopolds' Sand County farm was eighty acres. Eventually the family was able to purchase some more land. Today the Leopold Memorial Reserve consists of 1,500 acres that are owned severally. Access is, of necessity, strictly limited. The private landowners—Leopold's descendants and others—whose holdings together make up the Reserve are managing it as open space. They are attempting, by restoring natural processes, to favor indigenous vegetation in all the different ecosystems comprised in the Reserve. Such are the official purposes. From my standpoint of June-hedonism, what they were really doing was graciously maintaining a setting for a perfect summer day. I sat by the broad, shallow, but nonetheless formidable river secure that I would be able to remain comfortably out-of-doors, undisturbed, in a pleasant open privacy, knowing that this homey woodland is extensive, and pledged to be unspoiled. For restoration to take hold, protection must be ensured. The sense that this place would persist in its simple beauty, that one wouldn't have to dread the day that the Sande Countye Condominiums went on sale, lent an extraordinary peace and ease to the visit.

I sat in the shade of a basswood tree. The midday sun flooding the mowed opening nearby was almost too bright. Cottonwood fluff floated everywhere on the zephyrs, and olfactorily impaired mosquitoes braved the reek of the insect repellent I'd drenched myself in. Across the river on an island, a doe and her tiny fawn came out on the bank for a drink. Twenty yards in front of me on a snag just offshore, a couple of belted kingfishers perched, and made occasional chirring forays out across the satin surface of the water. The glossy stout oak leaves, stirred by the breeze, made a crisp, effervescent sound. The splendors of my day were small ones, and I found it poignant to think that a day of ease and rapt wandering in a lovely surround should be so exceedingly rare. Shouldn't every human being be able to have days like this, perfect summer days in a green world chiming with bird song, and bespeaking care for the land?

The Shack is the most unprepossessing little cabin imaginable, just weathered board and batten, a shake roof, a couple of Leopold benches flanking the door. Although the Shack itself, where the family bunked and cooked and stored their tools for weekend work, is now on the National Register of Historic Places, it is no sacrosanct museum, but a place still used for family weekends in the country. It possesses an ascetic coziness. There's nothing in it that's frivolous—camp chairs, a kettle on a tripod, kerosene lamps, a table and benches are about the only furnishings. The benches and mantelpiece were made of lumber that the kids salvaged from the river. Aldo scrounged the windows from the county dump. Another clever bit of foraging yielded snow fence to be used for box springs in the bunk beds, marsh hay for the mattresses. It's whitewashed inside with a few hand tools hanging on the walls, these not for decor. Approaching the Shack through a landscape that has been considerably healed by the Leopold family's efforts reminded me of attaining the Mona Lisa after threading the endless corridors of the Louvre. On arrival you discover an object that's small and unspectacular. You wonder whether it would mean anything without all the associations, but the richness of the associations supplies a tremendous aura. One of the realizations during this idyll was that this picturesque landscape, although not a howling wilderness, and not a grand range of light à la John Muir, but a quietly regenerating oak woodland, was nevertheless the setting in which the profound ecological truths of *A Sand County Almanac* were resolved into a comely form. Many of Leopold's ideas about beauty and land health were founded in wilderness, to be sure, but it was in this place that Leopold was able to see the land ethic whole. It was here, and not in virgin terrain, that Leopold articulated an ecological morality that addresses a range of relationships between humans and land, from exploration and discovery to healing.

During the first day of my visit to the Reserve I spent several hours with the manager, Matt Bremer, an articulate and amusing ex-Marine and veteran of the Wisconsin Conservation Corps. The Reserve includes an array of plant communities from floodplain forest to oak barrens to mesic forest to dry prairie to morainal prairie/savanna to old fields once cultivated for pasturage, now recovering. A nice assortment of "ecosystems small, distinct, identifiable," Bremer termed them. He explained that grazing had been the chief agricultural impact on the Reserve lands; only small plots had been tilled, and there was some clearing for wood. Therefore the tree component of the Reserve's landscape was considerably less affected than the herbs and grasses. The exotic Kentucky bluegrass withstood the intense grazing pressure, whereas the more delicate and diverse mix of endemic plants were all but eaten away.

As Bremer and I approached the Shack on a two-track, we passed a little grassland. This is known as the Shack Prairie, and was started by Aldo Leopold, a personal version of the prairie restoration work going on at the university's Arboretum in Madison. He transplanted mature prairie plants into some areas, scalped off the weedy sod and planted prairie seed in the bared earth in others, and also propagated prairie species in a nursery to plant in his plots. After twelve years of work, at the time of Leopold's death the prairie was flourishing. For twenty years afterward the Shack Prairie was neglected and lost much of its botanical richness. In 1971 Leopold's daughter Nina and her husband, Charles Bradley, began to give the prairie some skilled attention and it began making a comeback.

In 1992 as we surveyed it, Bremer pointed out that the lupine growing among the prairie plants was threatened by deer (which is to say it's being nibbled to death, a common problem in landscapes that have been fragmented and have lost their big predators). This led to an explanation of the Reserve's "Earn Your Buck" program, invitation-only "antlerless" deer hunts. ("Antlerless" must be a coinage arrived at to defuse sentimental objections to killing does.) Managing for "quality deer in quality habitat," as Bremer phrases it, culling the herd to sustainable levels, is directly in keeping with Aldo Leopold's reality-tested science of game management. Indeed, advocacy of doe hunting to trim deer populations to sustainable levels once landed Leopold in a peck of trouble with both the rod-and-gun crowd and urban bleeding-hearts.

The recovery of the lands around the Shack is under way, but far from complete. Because the southern Wisconsin landscape was fire-adapted (meaning that the plants were either tolerant of periodic burns, as many oak species are, or dependent on burns for access to light and a release of nutrients), during the periods when fires were prevented a dense underbrush was able to take hold, starving the ground-layer plants of light. This overgrowth hence reduced the overall diversity of the woodland floor. The present management philosophy at the Reserve includes the restoration of natural ecological processes by means of burning across ecotones, or areas where one plant community grades into another. They're trying to recreate the effect of lightning-set fires.

Man's second discovery of fire has apparently taken place in the minds of restorationists, and has provoked as much glee and wonder as it must have the first time around, maybe even the same sense of mastery. Bremer is very keen on "throwing" fires across ecotones. The resulting burns are ugly, but don't remain so. The plants that evolved here under conditions of occasional fire should thrive with this regime. Thus the fire-tolerant species, like white, black, and bur oaks, can be given room to breathe, and the invaders of the understory, like prickly ash and another prevalent weed called wood nettle, can be subdued. With periodic burning, the richer mix of wildflowers that would characterize the savanna community might have a chance to return and flourish in partial shade on their fair share of the soil's nutrients, and the little grassy patches—all called prairies—will get what they need also.

When I asked whether they are propagating and planting native species that might have characterized this kind of landscape, Bremer said, "I want to put the process back, not the community." The idea is that what should be there will survive. This led to a discussion of the restorationist's triage: "Should I be trying to recreate something that can't survive under present conditions or should I give Mother the process back and let her decide what it is that's going to survive?" In Bremer's opinion, it might not be worth his spending a whole lot of time on restoring a population of an endangered plant. Organizations like the Nature Conservancy are likely to do the work of keeping such species from extinction by preserving and protecting their habitat. Work at the Leopold Memorial Reserve has a different emphasis. There's no collection of biological rarities here. The Reserve's unique status—"It's private ground, it belongs to no program"—means that experimentation in restoration and management technique is, as it was in Aldo Leopold's day, what goes on here.

Some of the Leopoldian ethos is about taste—as in "a taste for country." Most of the plants reappearing in this place—the prairie and woodland wildflowers, known in the trade as forbs—are small and inconspicuous, unprepossessing, like the dainty penstemon and harebell I saw while walking along Pasque Flower ridge with Bremer. This suggests that our tastes have been not refined, but vulgarized by most horticulture—garish huge flowers on our bedding plants are to these native blossoms what Jimmy Swaggart is to, say, Black Elk. It takes a highly educated—either in vernacular or botanical terms—eye to see, much less identify, these delicate herbs belonging so well in their regenerating places.

In a warm and crisply detailed reminiscence of Aldo Leopold published in *The Land* magazine in 1948, Alfred G. Etter described a day with the Professor at the Shack. Etter wrote it in order to express "that deep gratitude which arises from the realization that he gave me across the depths of his spirit, and provided me with a rich portion of the confidence which overflowed in his life." Leopold had reason to be confident—he was well-born, well-educated, and well-loved all his life. And, as soon as he began to express his views within his profession, his words were well-attended. People looked to him to make a contribution of significant ideas, and so he did. Aldo Leopold was favored by fortune, and became a fine human being who was by today's standards remarkably unkinked. Much was given to him, and much he gave to the world, with a clarity and directness that are also remarkable by today's standards.

This confidence did not qualify Leopold's sense of the losses to the land. Etter recalled Leopold's heading, in the course of their morning walk, into a grove of old pines on a neighboring farm, and wrote that these "hundred-foot pines, scattered and clustered . . . demanded silence from us. . . . These were what the Professor wanted, if not for himself, then for those who would follow in his spirit. He was heartsick with fragments and remnants of the beauties he had once known. Somewhere he wanted them to be restored to inspire his successors with the knowledge of what could be, what once was." An ecological education can mean suffering as well as possible skill in healing. Its penalty, wrote Leopold, is that "one lives alone in a world of wounds."[4]

"The farm was a place to satisfy a craving for beauty and simplicity, and yet it was much more than that," wrote Etter. "It was where he worked out the subtle pattern that became his life. It was a place where he could gain information from the tallest tree or the most insignificant spring flower, from a casual squirrel or a cached fawn. Here he tried to piece together answers to the questions which Nature so often tempted him to solve. . . . Above all, this farm was a place where his children could learn the meaning of life and gain confidence in their ability to investigate small problems and discover things which no one knew."[5]

The greatness of Aldo Leopold's work, and that of his descendants, appears rooted in a genuinely happy family life. He was loved and favored by his parents, loved and was loved by his wife, and was loved and respected by his children. All in all, it seems that the joy in the Leopolds' life infused what they did on their land. They worked and played a lot, invented games, made music together. Aldo's daughter Nina Leopold Bradley talked about the restoration activity as being "so rewarding that it becomes a part of you." The ability to take pleasure in processes and relationships more than in substances and artifacts is damaged and underdeveloped in contemporary life, but restoring it is much of the essence of what we must be about. It seems to have been a bumper crop on the Sand County farm.

In late 1992 I gave a lecture at a Lutheran university in Southern California. As I recently had read and re-read Aldo Leopold's writing, acquainted myself with the outlines of his life, had the pleasure of a conversation with his daughter Nina, and the privilege of visiting the Sand County farm, my talk pivoted on Leopold's work. I asserted that no citizen of this country should be ignorant of *A Sand County Almanac*. (I am not alone in that belief. In various places around the United States are projects where funds have been raised to distribute copies of *A Sand County Almanac* as widely as Gideon Bibles.) After my talk I was a little dismayed to learn that *A Sand County Almanac* had been on a reading list of one of the university's classes, but that the students had requested that it be deleted. Apparently it has passed into a category of books too difficult for most college students to read, or worse, the students couldn't grasp its relevance to their lives. This is a bad sign. Simply living life in curious, respectful, helpful relation to some little extent of the land might suffice to get the basic idea. But you'd be missing the sense of history, the fresh, full-blown brilliance of ecological thinking on the page, and Leopold's extraordinary contribution to American prose. When I first began reading about the man, his life, and his work, I kept dissolving into tears. I wondered about this access of feeling. Perhaps one of the reasons was that, although the Iowan terrains in which Leopold pursued his youthful interest in natural history were hardly virgin, I realized that during his lifetime it still had been possible for him to visit extensive wilderness in the Southwest. In my lifetime the range of potential experiences in the land has dwindled as a result of what Leopold referred to as the "iron-heel" approach to living on Earth. It makes today's context for learning much thinner, biologically. Reading Leopold, who once wrote, "I am glad I shall never be young without wild country to be young in," I feel born too late.

My visit to the Leopold Memorial Reserve concluded with an interview with Nina Leopold Bradley. Nina received me at her home. The Bradley house is made, in part, of red pine logs thinned from the stands the Leopolds planted many years ago as a family—it's gracious and open, spare and comfortable, with a view of their garden and of a prairie beyond that that Nina and her husband, Charles, are working on. During our conversation, Nina told me about the prairie restoration methods they were using. Hanging from the rafters in the screened room where we talked were big paper grocery bags containing different prairie plants being dried for seed.

Nina Leopold Bradley has a huge handsome smile and resembles her father in roundness of face and straightness of nose. She's ebullient and vital, active in her concern. The only hint of her three-quarters of a century lived is disclosed when she gets up to answer the door and there's some stiffness in her walk. At one point during the afternoon, Nina showed me a phenological chart hanging on the hall wall—two big computer printouts based on about fifteen years' worth of data she and Charles have accumulated on the return and departure of migratory birds, the emergence of insects, and the flowering and going to seed of certain plants on the Reserve for each of what must be nearly 200 species. At seventy-five, Nina, part of the second generation of Leopolds to carry on restoration work, attests to the importance of staying with it. Here is what she said.

From the Shack you could look in all directions and see for miles. It was corn stubble and cockleburrs and the old chicken coop was full of manure. It was dismal; that was the end of the drought years. You know in 'Prairie Birthday' Dad writes, 'My own farm was selected for its lack of goodness and its lack of highway; indeed my whole neighborhood lies in the backwash of the River Progress.'

"He knew what he was getting, and he was doing it intentionally, I think, but he didn't sit down with us and say, 'Now look: This is what we're going to do.' But as soon as we got involved, first fixing up the Shack and then planting whatever needed to be planted, we all just fell right into it and loved it. When we asked the right questions, we really got a lot of information, but he wasn't one to be pointing things out all the time. It's interesting that we all took to it so quickly, especially Mother. And she didn't come from a family of outdoors people. She was a Spanish aristocrat from Santa Fe, New Mexico, and she loved it at the Shack!

"The way Dad bought the farm was also interesting. He had a good friend by the name of Ed Ochsner in Sauk City—they used to shoot bow and arrow together. Dad once said to Ed that he'd like to find a piece of property. So he and Ed looked around and he found this. Dad said, 'I'll take it.' He didn't shop around to see if he could find something that suited him better. Right away he got it.

"Five dollars an acre. He started out with 80 acres and at the time he died he had about 160. You know, we didn't have money in those days. It was just all you could do to pull together five dollars an acre.

"The whole area really was in the backwash of the River Progress. The barns were run down, and all the farms were in bad shape. Three crops of corn or wheat or whatever it was and you'd got all you could get out of the land. Everything that you can see now from the Shack has been planted, except a few oak trees along Levee Road right at the gate and one fairly good-sized hickory. There was nothing on our land but a few cedars on the sand hills along the fence row and a straight line of elms which since have all died. We even planted aspen. And sumac. And all kinds of stuff which now of course we're working hard to thin out.

"The pines Dad ordered from the Wisconsin Conservation Department, and of course he was interested primarily in the indigenous species. Every spring vacation we planted a thousand of each species; white, red, and jack pine. Dad came up with prairie seeds. I assume that he must have gotten them through his contact with the [University of Wisconsin at Madison] Arboretum. He would bring out transplants and he would take an area the size of that bed [4′ × 6′] and he'd clean it all up. If it was full of quack grass, he'd roll all that up and then hoe it and put in a few seeds. For years we'd see these square prairie plots all around the Shack area. Of course now they've all dissipated out; it was inoculating the area for prairie species.

"When Charlie and I came here, fifteen years ago, we started thinking in terms of prairie restoration. We had thinned the pines near the Shack, and we'd built this house and we had these enormous brush piles out in the prairie. We burned them and then we had a sterilized plot. To the east of the Shack there's a big area of lupine and puccoon and prairie dropseed. Those are our plots. They started out as, I suppose, plots of about twenty meters in diameter and now you can see all this stuff moving out.

"Nonnative plants are our very greatest problem. In restoration, I think, *that* is the $64 question. On this prairie out here, we used a different technique, one we learned by accident. We didn't use chemicals on it. Our pond is an old kettlehole, and it has silted in ever since glaciation, so we decided to dig it out. So we did and we had all these piles of guck. We spread it over about a three-acre area. Now the seed bank in that guck was probably aquatic. So we had no competition. [Meaning that the water plants whose seeds may have remained in the muck at the bottom of the pond couldn't have germinated under non-aquatic conditions; also that the layer of muck must have smothered the weed seeds that would have been latent in the soil of the three-acre plot.] And we have documented this plot over a period of years and we have no weed problem; there are practically no exotic species in it. The prairie across the road—we call it Two Bears prairie—we did treat with Roundup [an herbicide], weeded two or three times during the summer and then planted in the fall. No weed problem. My feeling is that getting rid of the exotics first is easier than trying to struggle with them later.

"We say this has inoculated the prairie to give it diversity. So we had originally about fifteen of these plots. Now we're actively working on about six of them. But as we get new species, we're adding them in those areas. And the diversity is definitely improved. We also are measuring all kinds of things, like the changes in the soil as a result of the restoration. Now this prairie out here [in front of the Bradley house] is now fourteen years old. In the first seven years of this prairie the organic matter in the soil went up 100 percent. So it is hopeful. By golly it is exciting to think that we can bring some of this destroyed land back to some kind of health.

"Our former manager had a black oak forest right near his house and the oak wilt invaded. Now, the oak wilt is a native bug and these trees were about seventy-five years old, maybe a hundred. The oak wilt killed off most of the forest and so he went in and cut all the trees down. Then as he was cleaning up he burned the whole area. And then the next year out came the most beautiful prairie you've ever seen. Big butterfly weed plants, not just little seedlings. Marvelous diversity! All the grasses, all the wonderful forbs. And we realized that that prairie had been sitting there for seventy-five to a hundred years. The oak wilt had invaded, the bugs had killed the black oak, and then it came back to prairie again. So then we had a graduate student work on this as a thesis and she went into the forest that had not yet died and she could see that a little bit of butterfly weed, just a few leaves, a little bit of prairie dropseed—all these species!—were sitting there. Even in the shaded area you could find just little pieces of these species. So that was a fun kind of a restoration, a great surprise.

"Lots of the animals are coming back. And birds. When Dad died, he figured that there may have been as few as 250 sandhill crane nests in Wisconsin. And now there are thousands and thousands of sandhill cranes." [Historian Susan Flader noted that Aldo Leopold thought that the species was virtually extinct. In "Marshland Elegy," he wrote, "The sadness discernible in some marshes arises, perhaps, from their once having harbored cranes. Now they stand humbled, adrift in history."⁶ He might be happy to know that there are now cranes nesting in the Reserve's restored marsh.]

Speaking of her father's depth of feeling, Nina continued. "Dad was very emotional. Things touched him *deeply*. I can tell you that there seldom was an evening dinner when I didn't see his eyes well up with tears. In his presence, there was never small talk. I mean never. People didn't do it!

"You know one of the last times that he ever spoke to a group, the newspaper comments were that it was an excellent paper, but poorly delivered. And I know in my mind why. I think he choked up. I've often wondered about that comment. I'm sure [the reporter] just missed the idea that it was an emotional experience. Isn't that strange? You'd think that somebody listening would understand.

"The Shack is frugality to the limit, but still it's very comfortable. Heat can be a problem. We used to say that we could be perfectly comfortable at the Shack until after the toothpaste froze. The Shack was nothing but it was everything. There was nothing elegant or beautiful about it but it was still comfortable. It was frugal and it was wonderful. I remember Dad used to say, 'We haven't room for anything extra so don't bring any stuff along (when we went to the Shack) that you don't need.' Well there were two things that we always took to the Shack that we didn't need. One was schoolbooks, which we never looked at but always brought, and the other was Dad's writing equipment which he always brought and never touched. He didn't write at the Shack. He thought. I think he did a lot of arranging of words and certainly a lot of thinking. But this was a place of action.

"In an undated paper, it must have been in the thirties or forties, Dad wrote, 'There are two things that interest me; the relation of people to each other, and the relation of people to land.' And as I think about it, both of these things were the main elements of the Shack. The relation of all of us to each other was just as important as our relation to the land. I can give you all kinds of examples of a spirit, of a feeling, of a harmony, of everybody pulling together, and of great fun and lots of humor. We developed all kinds of games. We would track in the snow and then backtrack and climb trees and try to fool the other people. We did broad jumps on the sand bar, everybody out there running and jumping, and we had a tree house, and we had a good time.

"Lots of music. Guitar-playing all night. Fairly recently I finally stumbled on that quotation: 'the relationship of people to each other.' At the Shack and on camping trips the men took care of the women. The men did the work. They did the cooking. And I can remember many a time Dad would bring Mother in, sit her down, give her a drink, make the fire, cut up the stuff for the stew or whatever it was. It was just a wonderful feeling. Everybody got it. All of us might decide to stay home [in Madison] on a weekend because there was something going on, but we would finally decide we had much more fun at the Shack and so we would end up there.

"I cannot remember a disparaging moment. That sounds like an exaggeration, but it really isn't. I can say that Mother's spirit is what made the thing work, the fact that she was having just as good a time. She was Catholic, but she didn't go to church when we went to the Shack. There was no question that it was more important to go to the Shack. And I loved that. I thought that really made a lot of sense.

"Mostly we were kids running around and exploring. We'd take off with an apple in the pocket. And then we really did range out, down to the next moraine down here and we got familiar with it. We would go upriver and check that out. We went over to the tamarack bog to the south.

"For the first time we, at least as kids, began to think of relationships: dry prairie versus wet prairie versus shade versus hot sun. It seems to me that restoration, and working with the land as intimately as that, is one of the most meaningful kinds of education in terms of land and all its concepts. And feeling for land as you see it change! By golly, you put these seeds in and you never expect anything to happen and then all of a sudden here you've got it!"

So here I am home in Michigan of an early afternoon, a bright, forty-degree day in late winter. Yesterday the south wind blew steadily, sculpting the snowdrift in front of my studio into sugary arabesques. The air is almost milky with what must be water vapor, rising off the liquefying snows. Today's temperature is causing the snow to become glazed on some surfaces, raddled and channeled on others, and diminished overall. It's also fostering the appearance of dense populations of snowfleas, little springtails that congregate like flecks of soot in shady meltwater depressions like the heelprints of my boots. Before this winter I hadn't remarked their presence. Now I wonder why they seem so prevalent.

Earlier in the day I looked out the upstairs window and surveyed the Scotch pine monoculture, wondering what was going on out back in consequence of the day's warmth and movement of water within melting snow. Then the following thought expressed itself, with an almost-audible whine: "I wish I had more ecologically interesting land to inhabit." Next, the thought that putting even a small footprint down in a patch of hardwoods (which I would consider more interesting) would only do harm. It would open up the canopy, move soil, introduce rapacious predators like my cats, and generally diminish the biota. Finally, I realized that the most conscionable way to possess and enjoy some ecologically interesting land is to "make" some—by restoration. Perhaps as a result of those days spent in the regenerating landscape at the Leopold Memorial Reserve, my own surroundings are beginning to tell more of a story, and its ending is no longer a foregone, or paltry, conclusion.

One finds this encouragement and cause for hope in Aldo Leopold's "The Conservation Ethic": "A rare bird or flower need remain no rarer than the people willing to venture their skill in *building it a habitat*.'"

Chapter 6

Learning Restoration

Although a day of rapturous attention on lush ground under open skies may be its own age entire, there are some settings that are more eloquent over the course of a year than in a day. Prairie is like that. Far subtler than a deciduous forest with its dramatic seasonal changes, a prairie's beauty shimmers around the seasons, with hundreds of species of plants—flowers and grasses—sprouting, ascending, blooming, and bearing seed—each in due time.

One June day in 1992, I was able to visit a created prairie at the University of Wisconsin's Arboretum at Madison.[1] Even in the company of a knowledgeable guide, I was afforded just a glimpse of the richness that was there, a richness that varies daily and is increasing through time. Even though June is a quiet time for forbs blossoming in prairies, there was enough spiderwort and puccoon and lupine, creamy false wild indigo, smooth phlox, scarlet painted cup, blue-eyed grass, wood betony, and golden alexanders that my eyes were quite feasted.

Later in the summer, when the prairie dock would attain its full blossoming height of six feet, and in August when the magenta pokers of prairie gayfeather would blaze away, and the arty, pale-green rattle-snake master assert itself, or in the late fall when the big and little blue-stem assumed a winy hue, only an apathetic soul could fail to be captivated, and the fires of spring, when the prairie is burned, and the ground appears ruined and blackened afterward, mark another change, another spectacle, another grassland season. The persistence on Earth of this prairie kind of beauty owes much to the work and study of the Arboretum's scientists and their students. The hope for restoration of many other kinds of ecosystems has been nurtured here as well.

Visiting the University of Wisconsin Arboretum at Madison seemed important to me, long before I'd ever heard of or contemplated writing about ecological restoration, or had learned of Aldo Leopold's role in the Arboretum's creation, which involved drawing up its wildlife management plan and envisioning its potential as a refuge and source of future wildlife for southern Wisconsin.[2] A trip to Madison in the summer of 1986 provided the occasion for a preliminary visit. I went on a Sunday when Arboretum offices were closed. No interpretive materials were available, so I curiously but cluelessly made my way along the firelane on the perimeter of the Arboretum's Curtis Prairie, guessing about the various woodland associations surrounding the prairie, and trying to tell the difference between the burned and unburned acres of grassland. According to the map, I must have wandered by an oak forest, a southern wet forest, a boreal forest, and a northern mesic forest. The Arboretum communities that went unseen by me on that initial hike are: savanna, open wetland, northern wet forest, southern mesic forest, and pine forest and barren. Even though I didn't make the full tour, I did manage a meaningful encounter with a woodchuck. Today I find it curious, maybe even fateful, that the place should have drawn me, willing but all unwitting.

The Arboretum is the place where ecological restoration first became a field of study (although, as we shall see, there had been earlier attempts at large-scale ecological restoration, and even the restoration of a local plant community in microcosm at Vassar College). The Arboretum's purpose, as Aldo Leopold put it in his dedicatory remarks

114

in 1934, was "to reconstruct, primarily for the use of the University, a sample of original Wisconsin." In the process of effecting that reconstruction with a degree of scientific rigor, and in the anticipation that the work could continue and be evaluated over time, the practice and theory of ecological restoration grew.

From its inception, the Arboretum was conceived of and is still used as a research facility. The University of Wisconsin at Madison has been host to a remarkable company of biologists who helped to plan and plant the Arboretum grounds; this Wisconsin-in-microcosm has been an epicenter for the development of the science and discipline of ecological restoration ever since. It is uniquely conceived, not least because it is local—to all of Wisconsin, that is—rather than being a planetary collection of exotica.

Its beginnings in the Great Depression, ironically, abetted the Arboretum's development. A crash in the real estate market allowed for the bargain acquisition of strategic tracts of land, increasing the Arboretum's size and the diversity of potential habitats. The Dust Bowl era, with its inarguable message of environmental devastation, had persuaded the Roosevelt administration of the urgency of land rehabilitation. The creation of the alphabet agencies, especially the CCC, with its widespread efforts to rehabilitate damaged land, provided a labor force for the Arboretum. In its earliest years of establishing prairies and forest plantations, the Arboretum was also a CCC camp, the only one on a university campus.

The places that the place comprises—the marshes, savannas, prairies, the deciduous and coniferous woodlands—are specimens, and small ones. To move about them is kaleidoscopic—one tours a condensed version of presettlement Wisconsin. Although the variety of woodlands, wetlands, and grasslands coheres, inasmuch as all of these kinds of forests and prairies might be visited on their home ground within a day's automobile ride, their extent is limited and the shifts between them are abrupt. It is a collection of ecological communities. It would take a great yearning to attain the feeling of being lost in and enfolded by an ecosystem in any one of them, the more so because the noise of the booming city of Madison, as concentrated by one of its freeways, is a constant presence. Yet parts of the 1,200-acre Arboretum, most notably the Greene Prairie, are idyllic and quite lovely.

At the dedication in 1934 Aldo Leopold said:

If civilization consists of coöperation with plants, animals, soil, and men, then a university which attempts to define that coöperation must have, for the use of its faculty and students, places which show what the land was, what it is, and what it ought to be. This Arboretum may be regarded as a place where, in the course of time, we will build up an exhibit of what was, as well as an exhibit of what ought to be. It is with this dim vision of its future destiny that we have dedicated the greater part of the Arboretum to a reconstruction of original Wisconsin, rather than to a "collection" of imported trees.

I am here to say that the invention of a harmonious relationship between men and land is a more exacting task than the invention of machines, and that its accomplishment is impossible without a visual knowledge of the land's history.

The time has come for science to busy itself with the earth itself. The first step is to reconstruct a sample of what we had to start with. That, in a nutshell, is the Arboretum.[3]

In addition to creating habitat for Wisconsin's plant communities—among them the vanishing prairies—the Arboretum created a habitat for a new kind of scientist—the restoration ecologist.

Wisconsin, being a midwestern place, had grasslands. The botanist Norman Fassett suggested that the restoration of a prairie community be attempted, and a tallgrass prairie plot soon became a focal point of the Arboretum. On account of his published study in prairie plant ecology, a newly minted Ph.D. named Ted Sperry was recruited, in 1935, to supervise the prairie restoration at the Arboretum. Sperry was due to visit the Arboretum within a few days of a visit I made there in the summer of 1992. His purpose was to update his observations of the plantings he had helped to establish on the Arboretum's first prairie restoration site—now called Curtis Prairie—nearly sixty years before. Sperry's interest in prairie plant ecology has remained acute since his pioneering days at "Camp Madison." His insight into the politics of that situation was also sharp, as I learned when I interviewed him at the Society for Ecological Restoration's annual meeting in Orlando, Florida, in 1991.

The octogenarian Sperry was a tight little package of sagacity, gallantry, and precision as he sat across the dinner table talking about his long career as a botanist, which began in 1927. Sperry's Ph.D. could not exempt him from the effects of the Great Depression, which was getting into full swing about then, so he found nonacademic work with the U.S. Forest Service in the upland oak hickory woods of the Shawnee National Forest in southern Illinois. "Any kind of job was something to celebrate," Sperry said. Wryly reflecting on the changing understanding of how best to manage ecosystems, Sperry remarked that "fire back in those days was the evil of all conservation, and 'Prevent forest fires,' the motto." One of Sperry's most significant contributions to the practice of prairie restoration would be experimenting with periodic burns.

Sperry said of the beginning of the prairie restoration at the Arboretum that "nobody knew how to make prairies then." The place where the Curtis Prairie would be established was derelict land "covered more by bluegrass and quack grass than anything else." To start the experiment, "We went out there and sort of messed up the field a bit." The prairie-makers dug up sods from a spot of undisturbed prairie on an old glacial gravel field twenty-five miles away and brought them in. "We scrounged for prairie plants where we could find them," said Sperry.

"We got a part of it planted as soon as the weather broke in the spring of 1936," he said. For the first two years it was "very unspectacular . . . and the weeds were still there. Once in a while there'd be one bright flower about as big as that blossom on the table." Before too long, though, enough plants came up that Sperry and his colleagues became convinced that something was happening.

In 1941, with the beginning of World War II, Sperry was drafted and the CCC's prairie planting work at the Arboretum terminated abruptly. The botanist John Curtis was able to carry on with the proj-

117

ect, however. Curtis, a plant ecologist, looms large in the Arboretum's history. His magnum opus, *The Vegetation of Wisconsin: An Ordination of Plant Communities*, published in 1959, remains the holy writ in discussions of the botany of upper midwestern ecosystems. Millions of plants were meticulously censused by Curtis, his students, and colleagues, to justify Curtis's schema. With graceful explanations of the formation and dynamics of dozens of different kinds of plant communities, *The Vegetation of Wisconsin* is a remarkable book, its lucid prose rising and blooming upon great moraines of statistics.

After the war, on one of Ted Sperry's earliest visits to the prairie, he played a role in the development of fire as a tool in prairie restoration and management.

In March 1946, after consulting with John Curtis, Sperry set out random strips across old plantings, and burned every other strip. In May, he said, the prairie plants in the burned strips were thriving. Blazing star was five times as abundant. He came back in August and knew by then that fire was definitely beneficial. "As far as I know, I'm the first one who ran this prairie fire experiment," he said.

The work in ecological restoration at the Arboretum in Madison was not without its historic precedent, although that history lay scattered and awaited the scholarly attention of contemporary students of land restoration as art and science.

In "Historical Perspectives on Designing with Nature," a paper given at the first annual conference of the Society for Ecological Restoration, Bob Grese, a professor of landscape architecture at the University of Michigan, describes the work of a number of early twentieth-century plant ecologists and landscape architects. Alarmed by the rate at which urbanization was destroying wild habitats, these thinkers and doers began to work with native plants and to educate the public about the need to preserve and foster native plant communities. Among the plant ecologists who took up the cause in the 1920s was Edith Roberts. Roberts had been a student of Henry Cowles of the University of Chicago, a pioneering theoretician of the succession of plant communities in landscapes. Roberts, writes Grese, "actively promoted the application of ecological theory in restoring plant communities." She became the chair of the Department of Plant Science at Vassar College and in 1920 began what she called "an out-of-doors botanical laboratory for experimental ecology."

The project was intended, she wrote, "to establish, on less than four acres of rough land, the plants native to Dutchess County, N.Y., in their correct associations, with the appropriate environmental factors of each association in this region." Roberts did this on a shoestring: "Using meager funds from the college, honoraria from occasional lectures, and hours of student labor and study, Roberts was able to establish representative samples of each of the major plant communities of Dutchess County with the majority of the plants raised from seed or spores," writes Grese.[4] Lamentably, although these plantings still exist, they have been neglected and so are overgrown.

Frederick Law Olmsted, the visionary nineteenth-century landscape architect, was instrumental in the design of several of the United States' great city parks (Central Park most notably), as well as in preserving areas like Yosemite Valley, Yellowstone, and Niagara Falls. Olmsted also "proposed what may well have been the first major attempt anywhere at ecological restoration as we know it today," according to Dave Egan, who is the associate editor of the Arboretum's biannual *Restoration and Management Notes*. In "Historic Initiatives in Ecological Restoration," Egan recounts how in 1818 Olmsted, working in concert with the city's engineer on a solution to the difficult problem of improving the miasmic conditions in Boston's Back Bay area, "emerged with a unique plan that combined sanitary engineering techniques with the cleansing properties and attractiveness of a tidal marsh."[5] The Back Bay fen, a restored tidal marsh effected by a complex system of sewage and stream diversions and by plantings of carefully selected species of salt marsh grasses, rushes, and sedges, had only a brief life, a decade or so, before other construction by the city eliminated it. Clearly, ecological restoration is an idea whose time has come a time or two before.

In June 1992, six years after my first visit to the Arboretum, I returned. This time I had both purpose and portfolio, recognizing the Arboretum's significance to the rapidly growing restoration movement, and wishing to take advantage of Dave Egan's good offer of help with this writing. I prevailed upon the affable Egan to act as docent for a brisk but illuminating tour of the Arboretum. We began at the forty-acre Greene Prairie, the masterpiece of botanist Henry Greene. "Greene taught in the botany department along with Curtis and Norman Fassett," said Egan. "Greene was the kind of guy who kept to himself. He took on this project on his own, didn't want anybody else to mess with it. They gave him the license to do it. He had enough respect within the community of his coworkers that they said, 'Sure. Go ahead.'"

The prairie, which Egan, a landscape architect, dotes on, is on a fairly propitious site within the Arboretum—it occupies land that once had been plowed and planted in corn, but whose wet wooded margins and nearby railroad rights-of-way harbored original prairie plants that suggested possibilities for Greene's composition. Egan helped me to see the restoration's painterly agreement with the parcel's subtle terrain. "The arrangements are in broad bands of color that show themselves at just the right times of the year," said Egan. "It's not patchy and small, it's patchy and big. It really spreads out over the forty acres. There's a lot of prairie dock [a plant with broad vertical leaves that sends up a tall stalk with numerous sunflowerlike blossoms]. Those are really something to see in midsummer. Then hidden within all of that are fringed and bottle gentians that you see when you part through all the taller grasses, and much smaller forbs, that flower when the grass is still short.

"When you start to look, you really begin to see the flow of the repeating forms, of the larger plants—like the leaves of the prairie dock and the bushy structure of the indigo. There are these larger, coarser things and then there are these little fine get-down-on-your-hands-and-knees-to-see-it things. And all of these forms are laid in these fine grasses, like bluestem and prairie dropseed in here." Could it be that Greene's sense of beauty accounted for the brilliance of his botanical praxis, producing a prairie that's an ongoing joy?

In a 1992 article titled "How Well Can We Do? Henry Greene's Remarkable Prairie," Virginia Kline, the Arboretum's ecologist and research program manager, evaluated the success of the Greene Prairie:

> The 20-ha Henry Greene Prairie at the University of Wisconsin–Madison Arboretum is one of the most successful prairie restorations anywhere, a fine example of "the best we can do so far." It is successful in terms of the usual objective criteria for prairie restorations: dominance by characteristic grasses, diversity of prairie forbs and grasses, little woody invasion, and few troublesome exotics. For many of the visitors following the trails, however, its success is measured in esthetic terms—the beauty of the prairie vistas, the colorful flowers in a background of grass, the remote location with its near-relief from highway noise. The songs of sedge wrens mingle with those of yellowthroats and goldfinches along the brushy edges, and an occasional redtailed hawk calls from above. To complete the sensory impact, mountain mint yields its pungent aroma in response to passing feet, and on a warm day late in summer the air is filled with the tantalizing fragrance of prairie dropseed.
>
> The success of this restoration is undoubtedly due in large part to the skill of Dr. Henry Greene, who selected and surveyed the site, and (at his own insistence) planted it almost single-handedly. He did most of the planting between 1945 and 1953, using seeds, seedlings and wild transplants. Greene was a botanist whose professional specialty was mycology, but he was an expert on prairies. Not only was he an excellent prairie taxonomist, but he knew the soil and moisture requirements for each species, and what combinations of species grew together naturally. Because of this he was able to do an unusually good job of placing each species where it would do well on the fledgling prairie. He took his time, kept the transplants watered until established, and meticulously recorded the location of each planting to facilitate later evaluation of its success.[6]

A plant ecologist with a doctorate in botany, Kline started her academic career in 1976 after raising a family of four children. Simply for some diversion during a difficult time she attended a natural history course in a local school. Soon she became involved as a naturalist in the school's outdoor education program. It wasn't long before she noticed that the decisions for forest management in that program were being made by credentialed men who didn't know as much as the naturalists (one of whom she'd become by then). Therefore, she said, "It was time to get some credentials." As she finished her degree she was asked to come to work at the Arboretum.

One of the more esoteric benefits of ecological restoration is that it affords real-life situations for field-testing ecological theory. Commenting on the value of restoration activity to the science of ecology, Kline said, "You find out more about community structure, action, and dynamics by trying to put it back together . . . [but] whatever you do you have more questions than when you started."

Of all the plant communities in the Arboretum, said Kline, "the savannas are what I'm most excited about right now," adding, "but that can change." (A savanna is a combination of grassland and forest in which trees are sparse enough that grasses and herbs are the dominant plants.) One of the savanna restorations under way is adjacent to Greene Prairie. Here, as in most savanna restoration, it seems, clearing out the exotic understory is a primary task, so Kline and her colleagues are having volunteers cut honeysuckle and buckthorn, then there'll be burning, and "first we're just going to watch what comes. Later we'll add species we feel are missing."

Savanna landscapes, she said, would come closest to looking like presettlement Dane County (where Madison is). She said they hope to have the common savanna birds, the redheaded woodpeckers, flickers, bluebirds, great crested flycatchers, as well as some prairie birds, "possibly even a meadowlark, if the prairie gets big enough."

To get an idea of why prairie restoration generates such interest, consider the biological diversity of your lawn. By contrast, a prairie may have as many as twelve or more species per square meter; at the Arboretum's prairies, there's a total of over 200 species—most of them planted. According to Kline, the health of the prairie restorations is good. "Curtis Prairie is in better shape than ever before," she said. The frequency of lots of species appears to be constantly increasing. "Each square foot out there has more occupants than it had before.

"I have marveled at the changes that have taken place."

In addition to serving as a setting for some of the first prairie restoration ever attempted, and as a laboratory where scientists for the last half-century have been experimenting with techniques in ecological restoration, the University of Wisconsin's Arboretum at Madison is also home to the headquarters of the Society for Ecological Restoration (SER). The Society, which was founded in 1987, is an interdisciplinary organization. Its members include various professional and amateur practitioners of ecological restoration. At its annual conferences, begun in 1989, restorationists of all stripes gather to deliver and listen to papers, to engage in discussions following on panels about some of the issues vexing restorationists, to bend elbows and visit with far-flung friends and colleagues at the evening social events, and to take field trips to restoration sites in various stages of progress. The biannual *Restoration and Management Notes*, edited at the University of Wisconsin Arboretum, serves as SER's unofficial journal.[7]

William R. Jordan III, *Restoration and Management Notes* editor, is a rightly prominent figure at the annual SER conferences, displaying his spellbinding creative intelligence at the podium, as well as in his editorials. I attended the 1990, 1991, and 1992 SER conferences and was treated to a series of intensive discussions of every restoration topic, from the picayune details of censusing plants, to the promising new relationship of humans to nature that restoration portends. Bill Jordan's thinking on that large subject particularly interested me so I sought an interview with him during my visit to the Arboretum. A terrifically stimulating conversation resulted.

I came here and I was not very well equipped to carry on, either by training or interest, a traditional nature education program," Jordan began. "So I said let's not just talk about nature, let's talk about the Arboretum. What's distinctive about the Arboretum? Of course it's this history of restoration. The next question is the one I've lived with ever since and that's 'So what?' Why is it important and what does it mean and what are the implications of the process of restoration?

"We're now beginning to realize that we have a task here, one for the good of the environment, where you might say the more people we can get out there, the better. It turns around the purely defensive posture where we've said the fewer people the better. I think this is a much more promising program for the conservation of ecosystems, because the ecosystems are now pretty much surrounded by and influenced by people. Throwing ourselves in front of the bulldozers and saying 'Let's minimize impact,' is next door to saying 'Keep out.' Of course that has its value, but it's not much of a relationship with an ecosystem.

"The first thing that occurs to the environmentalists is that this promise of restoration threatens to undermine their position. And that's true. But it can work both ways. And you do see it working out the first way in those mitigation projects in which the technology of restoration is being abused, and wetlands are being replaced with poorer quality wetlands; where nobody's concerned about the standards, that's a major issue. It worries both environmentalists and restorationists very much. The other side is Steve Packard's work [with the North Branch Prairie Restoration Project in Chicago], where suddenly the oak openings have a new lease on life in an intensely urban landscape where it's quite obvious that without a human commitment at the social and community and municipal level, those ecosystems are doomed. So it can work both ways. What technology can't?

"The relationship with nature which we have idealized is that of the observer or visitor. Where we've had trouble has been in finding a way for a person from this culture to have business in the ecosystem, in the way a Native American would. Native Americans didn't spend all their time tiptoeing through the woods. They inhabited the ecosystem and they interacted with it. The burning of the prairies is a good example. How do you do that with one of these classic ecosystems? Restoration's a way.

"There's a whole new message there, a positive message of entitlement to citizenship in the land community as a fully engaged animal interrelating with nature on its own terms, that is, through the exchange of goods and services—*economic* terms. I sometimes sound as though I think this is going to solve *all* our problems and in a sense I do because I think it's striking at the root of the problem.

"Acts transform people, and this act transforms people in a particular way. It gives them a basis for a commitment to the ecosystem. It's very real. People often say, we have to change the way everybody *thinks*. Well, my God, that's hard work! How do you do that? A very powerful way to do that is by engaging people in experiences. It's ritual we're talking about. Restoration is an excellent occasion for the evolution of a new ritual tradition.

"If you look on restoration as a sacrament of communion, is it not a kind of expiation, even a kind of penance? There's a self-abnegation involved which is penitential for the assertive Western mind. It is penance to sit down and patiently copy nature. Not artistically, nor creatively. It provides a balance to that intensely assertive, immensely creative vein in our culture. It's a very specific kind of penance, the giving up of just that thing which has led us to this trouble, so the recompense is in the same coin, in the same dimension.

"It isn't enough, having caused harm, or just having caused change, to say, 'We won't do it anymore.' There should be recompense, in kind. What do you do to recompense for causing change in the case of nature? What you do should be some rich, deeply-conceived act, carried out in terms that address the wrong.

"It's not quite right to call the use an organism makes of its environment *harm*. When we, for example, plow up a prairie to create a cornfield, that could be seen as a good thing, the first phase in a reciprocal relationship with nature. But if it's not matched by a giving-back it's no good, because that's the way ecological relationships are—they're not exchanges of X for X. They're exchanges of X for Y.

That's what individuality is all about, the distinctiveness of species. A tree takes in carbon dioxide and gives out oxygen.

"We take in the prairie, and what do we give back? One thing we can give back is an artificial prairie which is as like the original as we can make it. That's giving back in more than kind, because in making the prairie, we bring the prairie into our knowledge and consciousness, which gives it a kind of immortality. Insofar as it's incorporated into culture, the prairie is transcribed from its chemical-based, material existence into knowledge. Once the prairie has been restored, it's been understood in a new, more accurate way, and in an emotional and spiritual way. Then it takes on the kind of life that Shakespeare's sonnets have. One of the powerful ways of transcribing it from the one mode into the other is by restoration. So that's the 'more than kind.'

"In the process of restoring Curtis Prairie, John Curtis was given a motive for doing his meticulous survey of the vegetation of Wisconsin. It's a classic. Now we have that book [*The Vegetation of Wisconsin*] and, arguably, that's where the prairie is—that's its future.

"To come back to a more familiar and less ambitious scenario, let's just say that the future of the prairies of the Midwest will depend on our understanding of them, and on our ability to communicate that understanding to the next generation. That means getting it into the computer, getting it into the books. The restorationist defines quality in terms of resemblance to the classic ecosystem. The quality of the copies that we're able to have around here two hundred years from now is going to depend on that record.

"Having a landscape with lots of nature in it depends on finding a way to connect nature with culture," Jordan continued. "Restoration has a crucial component to bring to that relationship. Indigenous cultures generally tried to achieve some reciprocal relationship with nature, mediated into material and spiritual terms. Restoration, at the mechanical level, is the mechanical part of that reciprocal relationship.

"My vision, then, is that the society of restorationists will help bring that about, and that they will enter into collaborations with people from other cultures, with branches of the government, to get their craft up and going and to make their craft the occasion for public festival and celebration of our citizenship.

"Here's one vision for the future: you're in the Midwest and it's October, and you're driving into northern Indiana and there's a big sign, and it's on the radio and it says the interstate will be closed for the next two days because everybody is out burning the prairies on the roadside. You can drive out there but the speed limit's 15 miles an hour. You park by the roadside and everybody's burning the prairie. There's song and dance and festival. The big trucks are parked by the side of the road and the truck drivers are out burning the prairie. Those prairies, in addition to their festival function along the roadside, would be integrated into agriculture as part of a crop-cycling process. We'd take and give. You'd plow up the prairie and plant corn for a while and then replant the prairie and replenish the soil. So that again, twenty percent of Indiana is always tallgrass prairie, but it's shifting around."

Articulating his insight as to our species' way of being in our ecosystems, Jordan said, "The human being is essentially a social creature and the relationship between humans and nature has to be negotiated on a community-to-community basis. It can't be just Thoreau at Walden, or John Muir up in the Sierra all by himself. That's not in any way to reduce the importance of the individual's contact with nature. It is to say that outside the human community, it's kind of a loss: the full relationship can't be negotiated. How can an individual negotiate a relationship with a community? One has to think in terms of not just private ritual but public liturgy. Restoration must become a community event in a way that backpacking can't. Restoration works well at the community level. You have in the field a human community which can really begin to negotiate a relationship with the biotic community; all the resources of social life are there, which can enable a person to negotiate this very problematic relationship."

Part of the problem of that relationship has been intellectual, an unfortunate tendency to objectification. Jordan commented that "science has these two faces; the 'poke in there and get out the secrets' kind of science, and then the other, which is more like the science of ethology. I don't know whether it's fair to segregate the sciences, but certain sciences—and ecology is surely one of them—are sciences more of the whole. They treat nature more as a subject than as an object. That's clearly the case with restoration. Ecology is the core science and is being taken seriously as the basis for a healing art. Nature is subject here, not object. The humility comes as the commitment to restoration is a laying aside of one's tastes and imagination, one's instinct to create a bigger, better ecosystem—one that makes more beans per acre—all of which we love to do as a species, and certainly as a culture. Restorationists take a kind of vow of self-abnegation, self-effacement, like a vow of celibacy. That, I think, is pretty humble. The restorationist presumes to copy nature. And that is presumptuous but it's a good thing to try to do in all humility because it yields lots of lessons.

"Over the past generations, environmentalism has, in its need to protect what remains, often spoken of ecosystems as fragile and irreplaceable. In a general sort of way, that's right. *Vulnerable* might be a better word. Fragile suggests a wineglass or a windowpane. We don't talk about breaking living things, we talk about wounding. Unless you specify what ecosystems are fragile or vulnerable *to*, what do I do about it? If the idea is to keep me out, it's enough to tell me that they're vulnerable to everything, but that's a pretty minimalist basis for a relationship with the ecosystem. By the same token, saying that ecosystems are irreplaceable is a very vague expression. It's strictly true in the sense that if you change it, you'll never get it back exactly the way it was before. But ecosystems are constantly changing and they're constantly replacing themselves. Yes, extinction is forever, and the chances of getting the woolly mammoth back are slight. We'd be very smart to regard endangered species as irreplaceable. But if you insist that ecosystems are irreplaceable, you preclude any hope of restoration.

"Exact restoration is impossible. So is preservation. So let's get on with the conversation. What we're drawn into is engagement with a system where we accept responsibility for our influence on it, even when we don't fully understand that influence. That's one of the wonderful things that emerges from this story. We have all these influences on the ecosystem which are not only inadvertent, but *invisible* to us. What I'm driving at is that it's the commitment to restore the ecosystem that forces us to explore all that we've done to the system, and to uncover all of these hidden, unseen, or unrealized influences. *That's* how we get to know who we are in relationship to that system. *That* experience generates an ecological definition of who we are."

At the Arboretum, and throughout the Society for Ecological Restoration, they talk about restoration and management. The "management" part of that phrase is the measure of the problem, at least as much as it is an element of the solution. Even after hearing Bill Jordan's response to that concern, I found myself a little uneasy still with the idea that an ecosystem, which is ultimately unknowable in all its detail (albeit perhaps understandable in principle), might be amenable to management by humans. Yet as the situation and stance of the Arboretum make clear, restoration and management of ecosystems is an experimental thing—ideas are being tried; and whereas the experiments haven't been running nearly as long as natural selection, a great many such efforts are clearly necessary. Given the generally perturbed state of the atmosphere, hydrosphere, and biosphere, some conscious intention and objective practice of healing damaged land may be our last best hope. "These questions are of national importance," said Aldo Leopold in his dedicatory remarks in 1934. "They determine the future habitability of the earth, materially and spiritually."

Chapter 7

Prairie University

I want it to be next spring already," says John Balaban. Balaban is one of a network of 3,800 volunteer stewards, laypeople who donate their time to the work of restoring prairie and savanna ecosystems in northern Illinoisan nature preserves. Many of these preserves are suburban and cluster around the North Branch of the Chicago River, within fifteen to twenty miles from the downtown Loop. At the height of summer, Balaban wants it to be "next spring already" because with every passing year, more is learned about the art and science of repairing damaged landscapes, and for many years John Balaban and Jane, his wife, have been ace practitioners.

Jane Balaban, a hospital administrator, is bright with enthusiasm for stewardship and restoration as she welcomes me to her comfortable home and offers me some lunch. It's August 1991. I have come here to rendezvous with Steve Packard, Science Director of the Illinois chapter of the Nature Conservancy and a prime mover of the North Branch Prairie Restoration Project and the Volunteer Stewardship Network.[1] He has kindly consented to take me on a field trip to some of the sites the volunteers are working on within the Forest Preserve

District of Cook County, 67,000 acres of mostly wild land scattered north and westward through the Chicago metropolis. (Despite the sylvan name, the ecosystems contained in these preserves are not only forests, but grasslands and savannas, which the early settlers called oak openings.)

Fire is the elemental fact in the evolution and dynamic equilibrium of these landscapes, as it is in Leopold's southern Wisconsin country, and all grassland ecosystems. Left to its own vegetal devices brush would shade out the oaks, and the many native grasses, sedges, rushes, and forbs that characterize these rich tapestries of life in which no single color or thread, no one species, predominates.

The midwestern tallgrass prairie all but vanished, overturned by the steel-bladed plow in favor of the relentless monotony of agriculture, which amounts to planting lots of the same damn thing—corn in Illinois, usually—and then attempting to protect it from all the plagues that even-aged monocultures are prey to. Sodbusters and years later the advance of suburbia made wild fires, and hence the prairie and oak openings that depended on them, even more a thing of the past. Benign neglect will not suffice to keep these precious few remnant ecosystems alive. Fires must be lit and skillfully managed, and many other tasks performed as well.

Prairie and savanna restoration work is terrifically labor-intensive, requiring thousands of hours of work to remove exotic species of plants from the areas to be restored, to lop the light-hogging buckthorn and ash saplings, to scythe weeds, to burn the leaf litter and grasses, and to gather, thresh, label, store, and then sow and rake seed from hundreds of different varieties of rare plants. Volunteers are *sine qua non.*

The "idea that these sites can't exist without our help anymore"—as John Balaban put it, "you can't just preserve something by building a fence around it," because of "how dependent that structure [of the ecosystem] is on our interference"—convinced the Balabans of the need for their stewardship as it has hundreds of other volunteers.

The North Branch Prairie Restoration Project is a remarkable mobilization of human energy and is instructing a sizable group of amateurs in the Chicago megalopolis in some of the finer points of prairie and savanna ecology, preserve management, field botany, plant conservation, and horticultural technique. To enhance and embellish this learning, volunteers coordinated by Laurel Ross, who like Packard works for the Illinois chapter of the Nature Conservancy, publish a thrice-yearly catalogue for "Prairie University." This university-without-walls has a curriculum consisting of courses available at educational institutions throughout the region as well as workshops, seminars, and field trips offered by various museums and learned societies and some by the volunteers themselves. Among the scores of offerings listed are: aerial photography, art/nature, biology, birds, botany, chemistry of the environment, controlled burns, ecology, endangered species, entomology, environmental science, evolution, field methods, gardening and landscaping, geography, geology, mapping, mushrooms, trees, wetlands, and zoology. The City of Chicago's motto is *"Urbs in Horto."* The Prairie University motto is *"Discere in Horto."* It means "learn in the garden" (also, but unintentionally, "learn by exhorting").[2] Paging through the Prairie University catalogue is a cheering experience, evidence of the hankering of so many members of the human community to exchange knowledge of natural history and promote ecological literacy.

"Know the plants" may be coequal with the injunction to know thyself as a fundamental responsibility of *Homo sapiens*. Plants tell us where we are, are the basis of our sustenance and our atmosphere. They make the life of the land. Ecological restoration is very much about the reinstatement of native plant populations and communities, and letting loose the processes they require for their evolution.

It is also important, in becoming native again, ourselves, to know the animals. And, since *Homo sapiens* are calling the shots, working with the dynamics of our own species is a fundamental aspect of restoration. In addition to all the botanical and occasional entomological savvy that informs the work on the North Branch Prairie Restoration Project, there is a good deal of human understanding. Not for nothing was Steve Packard a student of social anthropology. He and his prairie project colleagues work in an urban setting where people are over-whelmingly the dominant species.

What Steve Packard is famous, even notorious (among pedigreed plant systematists), for is his assertion that there is a distinct but for-gotten community of grasses, shrubs, and wildflowers existing in oak savanna. Before Packard's savanna hypothesis, the idea was, roughly, that what hadn't been oak forest had been tallgrass prairie. Therefore prairie was what it would make sense to restore to in these open wood-land places. The problem was that the prairie restoration didn't take. In areas of the Cook County forest preserves where the great old oaks lingered, even after clearing out the underbrush and planting prairie species in the newly opened but partly shaded ground, volunteers ob-served that the presumed natives failed to thrive. The prairie plants couldn't seem to take hold and displace the invading woody species. At the same time, what Packard referred to as "a few oddball species of plants" characteristic of neither prairie nor forest kept popping up in the sites.

By some masterful sleuthing Steve Packard began to speculate about what community of vegetation might have flourished originally under scattered tree canopy. Through taxonomic and historical re-search, he assembled the identities of the few "oddball species" and many other plants in the oak savanna complex. Eventually he came up with a list of plants that turned out to be comprised of woodland, not prairie, grasses, and many herbs. Scouts began to locate other small remnants of savanna communities within the vicinity of the preserves and to gather seed and propagate plants for savanna restoration work.

Before we set out for the Somme Woods Preserve, northernmost of the North Branch project sites, and Packard's "favorite place in the world," he took me around to the side of the Balabans' house to show me a little garden where the couple was growing some of the rare prai-rie and savanna plants that they and other volunteers are restoring to these sites. Household cultivation of these scarce plants weaves hu-mans together with nature, and the developed landscape with the re-flourishing landscape of the preserves.

Over the long run this learning in restoration—along with the landscaping-with-native-plants practice increasingly promoted by garden clubs, enlightened landscape architects across the country, and networkers like Louise Lacey (editor and publisher of the northern Californian *Growing Native*)—could enliven, maybe even transform, the biological monotony and heterogeneity of human settlements.[3] Such work, and the propagation going on in the Balabans' garden, and hundreds like it, is clearly reinhabitory.

In the Balabans' garden several members of the preserve ecosystems are making themselves right at home to the point of reproducing. "The tomatoes are over there," says Packard. "The endangered small sundrops are here." More than 100 members of the Volunteer Stewardship Network are, like the Balabans, gardening for seed, most of which is used in the preserves (although there is some seed sharing with other worthy restoration projects nearby). Wild seed is gathered from spontaneous prairie and savanna plant populations within a close radius of the preserves. It goes to the Chicago Botanic Garden for propagation, and seed yielded by these plants also is returned to the preserves, or shared. "The Garden doesn't have the resources to get the seed that we do, because we have so many people who are out crawling around who can recognize the rare plants," explains Packard. Meanwhile, the Botanic Garden has facilities such as greenhouses and mist tables that are crucial to propagating.

Packard points to a particularly strange and wonderful prairie plant in the Balabans' garden—the pale Indian plantain, which lurks at ground level for years in the form of a modest rosette of leaves, sending down its root system, all unseen. Then the plantain bolts, thrusting up a six-foot stalk with clusters of pale fragrant flowers which attract quite an amazing convention of flying insects. Among them was a raven-colored wasp, a midnight jewel prowling and flicking its way among the creamy little blossoms.

It is possible to grow the rare prairie and savanna species in plain old backyards because such domestic landscapes reproduce, albeit in extremely simplified form, the basic structure of the savanna: "Grass, and trees and flowers, partial sun and partial shade—that's what people like," says Packard. Indeed savannas are thought to be the landscape in which we *Homo sapiens* differentiated from our tree-swinging ancestors. As climates got drier (as they are wont to do from time to geologic time, thus forcing evolution's hand to new and interesting adaptations), trees became too widely scattered for brachiation. Eventually some -pithecus or other confronted the necessity of walking upright in a parklike world of grasses, trees, and flowers, partial sun and partial shade, and found it good.

Steve Packard apparently was born to be an ecological restorationist. Just as some kids are always going to be firefighters or doctors or *danseurs*, he was always going to be a restorationist, although it took him a while to discover how that was the case. Since junior high school he has found ways to see places become more like themselves, to regain their true identity through time.

"I used to make terrariums when I was a kid," Packard said later in our encounter. He told stories of leading his teachers out through the Massachusetts woods in the winter and dowsing for specific plants under the snow, and then going on to teach others about making terraria. "The more germane thing," Packard later added, "is that I planted native strains of pines and hemlocks in my behind-the-backyard 'bird sanctuary,' fiercely resisting my dad's pressure to use the 'improved' strains you can buy in a nursery."

Highly educated (Harvard, but reticent about it), Steve Packard now goes about retrieving lost landscapes through dirty-fingernails work mixed with long-term contemplation and cogitation, as absorbed as a chess master. During my day visiting restoration sites with him and later over dinner, Packard's conversation evinced a curious amalgam of self-effacement and pride. He credits his initial learning of plants to Roger Tory Peterson and Margaret McKenny's *A Field Guide to Wildflowers of Northeastern and North-central North America*. The book, Packard says, put in the layperson's hands a tool for calling the plants by their proper names without having to do the taxonomic trudging. (He has long since learned to wield a mean taxonomic key, sussing out even such cryptic native flora as grasses and sedges.)

He had a botanical epiphany in Chicago in 1975 when he was transfixed by the sight of a beautiful, fresh white flower, *Lychnis alba*, the evening campion, growing on some desolate ground behind a factory. At about the same time he read that someone had discovered the last prairie. "When I traveled down to see it, it seemed pretty magical in some ways . . . in some ways it seemed like a lot of weeds." The magic of it won out and thus began a vocation in prairie and savanna restoration—one that over time would come to entail some great battles with a lot of weeds.

The visionary prairie-phile landscape architect Jens Jensen played a leading role in the establishment of the Forest Preserve District of Cook County. Jensen practiced in the region around Chicago in the early twentieth century. In addition to being one of the first to celebrate native ecosystems in public parks, Jensen promoted volunteer participation in the work on certain of his projects. But neglect of the necessary management practices led to the preserves' degeneration. Enter the North Branch Prairie Restoration Project, whose *raison d'être*, according to its Mission Statement, is "to assist the Forest Preserve District of Cook County and other agencies in protecting and restoring native Illinois ecosystems."

The preserves were established sixty years before the Volunteer Stewardship Network got started. There have been no fires in them since the mid-nineteenth century. In their ecologically degenerate state, much of the land in the preserves had been abused and used the way vacant lots generally are—as kids' rendezvous, party spots, offhand trash heaps. Removal of soggy car seats was an early step in the restoration process. The absence of fire had allowed a proliferation of brush, particularly the European buckthorn, which shades out the oak seedlings and stifles the lower branches of the centuries-old oaks that remain, talismans of a time before Chicago's settlement by Europeans. The restoration work on these preserves includes bringing the brush to heel, then planting seeds of the native climax vegetation. The theory is that the former climax community will succeed, in both the vernacular and ecological senses of that term. A climax community maintains conditions—soil texture and nutrients, shade, and biotic richness—that favor its continuation. Barring disturbance, climax communities are very stable through time. Nature being the prankster she is, however, disturbance inevitably occurs, and in the disturbed areas successional processes are set in motion. Different groups of plants are adapted to take advantage of the set of conditions prevalent at the moment, each through its life cycle changing those conditions toward climax. In the succession to forest, for instance, the movement is from open, sun-drenched poor soil through ephemeral light-hungry species to long-lived, shade-tolerant trees that will form a canopy with moderated conditions of temperature and humidity below. Grassland communities and oak savannas seem to create the conditions required to invite periodic burns which kill the fire-intolerant brush species that would shade out their light-loving members. Notes Packard: "Prairie, savanna, and oak woodland might all be called 'fire climax' communities. Many other ancient communities are also known now to require occasional disturbance (fire, flood, blow-down, disease). Natural disturbance functions differently from the catastrophic disturbance that humans often wreak."

From the roadway, there's no clue to the extraordinary botanical goings-on at the Somme Woods Preserve. The crafty stewards have left a hedge of buckthorn around the perimeter of the woods, an impenetrable barrier to deter the insensitive, or the destructive visitor (the sort who, until recently, had come to drink and neck and hare around on dirt bikes). On arrival we threaded through a small opening. Packard charged me to behold "a sick, miserable ecosystem that used to be oak savanna." At the beginning this place must have looked horribly unpromising. "There were pockets of good stuff and dribs and drabs of this and that species, in dwindling numbers, dodging the disturbances," notes Packard. "Or perhaps even dependent on those disturbances—in the absence of fire." He pointed to a stand of locust trees that the volunteers had girdled, killing them to let the light shine on the tree species that belong there—bur oaks and white oaks. He pointed out a young oak that was saved from dark demise under a dense canopy. He explained the hallmarks of fire-adapted trees: their corky bark, their fat twigs, their ability to resprout.

Our passing from dark to light to dappled light was the subtext for the lesson that Packard imparted as we walked. We moved first through dark, sterile thickets, with their bare, caked mud floors, then prairies, and then the oak savannas that Steve Packard considers the epitome of desirable terrain. The contrast between the richness of the communities that the network has reestablished in the savannas and the nullity of the ground under the buckthorn is truly dramatic. I had no idea there were so many different kinds of grasses. In the course of our walk we remarked perhaps two dozen different species and many sedges as well.

This list of names from my notes as we toured the Somme Woods Preserve may give the reader some fractional sense of all that's there: satin grass, slender wheat grass, nodding wild onion, *Carex aurea*, mad dog skullcap, Virginia wild rye, bottlebrush grass, Canada wild rye, wood reed, woodland nodding fescue, a monarda, a vervain, ironweed, rice cutgrass, Illinois rose, hop sedge, bulrush, sweet black-eyed Susan, Kalm's brome, hill's oak, great Saint-John's-wort, mullein foxglove, sneezeweed, *Cirsium discolor* (pasture thistle), prairie drop-seed, Riddell's goldenrod, meadow rue, Indian hemp. This random list does not include the really obvious prairie plants that any school kid ought to know, like big bluestem, blazing star, leadplant, rattle-snake master, prairie dock, and compass plant.

There's a level of perception at which vegetation is generic. For most of us the landscape either has green things growing on it or not. It takes a while to begin to learn the difference between old-growth and second- and third-growth forest, or the difference between open land populated by exotic species—old fields—and open land like the tallgrass prairie that is richly populated by a diversity of species. Accustomed as we are to having to visit zoos to see exotic animals and botanic gardens to see rare plants, it's quite a wonder to be able to look out across a pelage of dozens of rare, and some endangered, species, all growing together and turning their faces to the same sky that arched over their ancestors' arrival 10,000 years ago on plains recently bared by the glaciers' retreat. At the restoration sites in the Cook County forest preserves, a respectful visitor can tread a narrow footpath past healthy populations of plants that, until a few years ago, were rare waifs and getting mighty lonely.

Splendid as it all seemed to me, the drought that summer of 1991 was affecting the physiognomy of the site, and Packard fretted about the fact that the restorations weren't putting their best face forward. Under normal rainfall conditions, there would have been many more flowerings, he said. A climax ecosystem such as this, though, has evolved to deal with extreme variations in weather. If the native community can regain its rootholds in these places, runs the theory of successional restoration, they won't all look great every year but they'll outcompete the invading exotics. It's the diverse community against the monoculture.

The thing that is changed forever (or for a long time into the future, anyway) about the life of this renewed ecosystem is that the fire it needs must be set and tended by knowledgeable humans. It calls for modern urban slash-and-burn horticulture. The volunteers cut out the brush to roll back the shade. Once there's enough fuel on the ground—leaf litter from the native plants—then fires can burn now and again and maintain the vegetation complex. Prior to settlement most of these fires were set by indigenous people to maintain clearings for game. Other wildfires would have been lightning-set. The occurrence and intensity of the fires was as irregular as rainfall. Different fires burn at different temperatures; different plants succumb to different fires burning at different temperatures. The fact that some plants survive and some don't results in a mosaic pattern in the vegetation rather than in a wall-to-wall carpet of repetitious motif.

There was a moment of pure bliss for me in this postage stamp of healthy land. We walked in an island of wildness that had a surf of automobile traffic noise beating around its shores. But there were enough cicadas stridulating to give the traffic some stiff competition. The weather couldn't have been improved on, the sky blue and bright with high clouds, twinkling with butterflies. Butterflies like these places. Packard remarked that even if you didn't like plants, this work would be worth it for the butterflies. We watched a tiger swallowtail and an ebony-colored black swallowtail, or it may have been a mourning cloak of similar size, dancing a *pas de deux*, or they may have been contending for territory across a prairie opening. There was just something about the light. And about the excitement of being among so many different kinds of plants, of feasting the eye and the subtler senses on bountiful detail. Always something fresh to point to and ask, "What's that?" It was a great privilege to be with someone who knew the answer, who may in fact have caused the cluster of false indigo or Indian grass to be growing there at all.

The Somme Woods site which we toured is, as is any restoration site, a lab. It's a story featuring the uneasy waltz of Science and Nature (Science all too often showing up for the dance in steel-toed work boots while Nature goes barefoot). Restoration ecology is experimental science, a science of love and altruism. In its attempts to reverse the processes of ecosystem degradation it runs exactly counter to the market system, to land speculation, to the whole cultural attitude of regarding the Earth as commodity rather than community. It is a soft-souled science.

At one point in the ramble I asked Packard what he felt in such places. If he'd been there by himself what would he be feeling? Would he feel a sense of satisfaction? After some thought he said he was wary of taking personal credit for this. He likes almost to forget that he and others have had anything to do with this place, this recovering ecosystem. It is as it should be, and that's the point. He talked about the satisfaction that comes of being a part of something good and useful. He made an analogy with taking a child to do something fun. The child mayn't necessarily thank you, but being able to witness his or her enjoyment is reward enough.

Packard talks about organizing the "generous impulses that people have toward their world." As an organizer his method is to say, "Here's the necessary support, here's the necessary freedom and authority: go ahead." It amounts to a deft liberation of energy, akin to the emancipation that the burning and girdling of underbrush affords the hitherto scarce plants like the sweet black-eyed Susan which, said Packard, "needed the same release that the people needed."

An ecological community is founded on plants, the primary producers, with intertwined loops of consumers—herbivores, carnivores, and omnivores—relating hungrily in the midst of the vegetation. An ultimate index of health is the presence of big, wide-ranging predators. Because the preserves are so constricted and isolated in size, the likelihood of reinstating larger creatures seems slight just now. But very large creatures once were integral to these ecosystems. Chicago was a place where the buffalo roamed and what became Vincennes Avenue was once a bison track. However, the Somme Woods Preserve, as Packard told me that day, is host to red-tailed hawk, sparrow hawk, and the rare Cooper's hawk. Evidence that a goshawk wintered here is provided by the piles of pigeon feathers found all over the preserve. So there are avian predators and coyotes that have returned, but no wolves or bison are expected for a while. The question is whether these restorations will be sustainable indefinitely in absence of the larger animals that were once a part of these communities. On the savannas and prairies humans have to stand in now for the forces of Nature, to play the part of lightning storms, of buffalo, and of wolves and cougars, as they cull the deer.

White-tailed deer have become so numerous in the North Branch prairies that their grazing threatens to do irreversible harm to these painstakingly restored ecosystems. The decision to kill as many of the deer as possible triggered a controversy among the Chicagoans paying attention to such things, and prompted some opposition by animal-rights activists. Laurel Ross, who coordinates the Volunteer Stewardship Network, engaged the subject in her writings:

> These caring and well-meaning people speak loudly and passionately against the culling of deer, as if saving some individuals should take precedence over saving the precious and irreplaceable system which supports them. It is important for us not to lose sight of the fact that *we*

are animal rights activists. We put our time and our energy into restoring and preserving habitat so that hundreds of species of animals and plants may thrive.

When newcomers to the Prairie Project challenge the seemingly brutal methods used to control brush and pest plant species, it is explained how and why this is critical to the recovery of an ailing ecosystem.

Deer control parallels other management techniques. Because we know fire is essential but cannot live with uncontrolled wild fire in a populated area, we have substituted safe, controlled burning. Nature provided wolves, bears, mountain lions and people to keep deer populations in balance. If we cannot tolerate these large predators in our cities, then we owe our natural areas a substitute they can live with.[4]

So how does the Forest Preserve District actually do the dire deed? A trench is dug to hide a net. Bait is set out for the deer. When they are gathered in sufficient numbers, they trigger the flinging, by small rockets, of a net over the herd. Then the deer are shot point-blank. Their meat is distributed to various charities.

Counterbalancing the drama of sweat, smoke, and blood being shed in controlling superabundant species, an aspect of savanna restoration, which is in some ways the essence of the whole thing, is seed collection. There's no catalog whence you can order seed envelopes for the scores of woodland species the project seeks to reestablish in their rightful venues. (Besides, even if such could be purchased, it would lack the critical attribute of ecotypy—genetic adaptation to a certain locale.)

Seed collection is the endeavor Laurel Ross superintends as a volunteer. On a visit in the fall of 1992 to a different restoration site in the forest preserves—the Bunker Hill Prairie—I tagged along during a field exercise. During this consultation with restorationist Tom Vanderpoel, Ross was also reflexively scouting seeds. As we walked along in the company of John and Jane Balaban and several other volunteers interested in refining their understanding of, and planning for, the ongoing restoration and management of this preserve, Ross noticed smilax seeds ripening. In another place, she noticed that a certain sedge's seeds were ripe and produced a half-dozen plastic bags into which we were directed to gather the harvest.

Plants are impressively protean, taking many different forms, shapes, and sizes in the course of their lives. Consider, for instance, the difference between an acorn and an oak. An untrained observer (such as myself) may see the same plant in a half a dozen different places, having made its adaptation to a half a dozen different sets of circumstances—to peculiarities of microhabitat or to passage of time—and be unable to perceive that they're all individuals of the same species. Thus the ability to recognize hundreds of different plant species, and in the various stages of their life cycle, is real *savoir vert*.

"It's like people," Ross says. "You have to know them really well, so that you can know them walking away from you, when they've gained ten pounds, when they're half asleep."

At length we came to a setting that Vanderpoel figured would favor savanna grasses and forbs. "You can really bomb away with the good stuff here," he said, meaning seed it in heavily. Packard, alluding to the preciousness of such seeds, dryly pointed out that one wouldn't be strewing megabombs but thimblefuls. (Ross later noted that she had counted seeds once or twice to remind herself that there can be many thousands in a thimbleful.) Pinch by pinch, inch by inch, is how this work actually goes.

Further on, Packard pointed out a stand of box elder and other woody fellows that have snuck into what fifteen years ago, when the North Branch Prairie Restoration Project was started, was an open field. "You just have to go toe-to-toe with brush," said Vanderpoel. "You've got to be fanatic." Cut it down, hit every stump with herbicide to prevent its resprouting was the advice. Yes, restorationists do use herbicides, albeit very selectively. They are just a tad defensive when questioned about the practice, which seems to be here to stay. In the emerging armamentarium of ecological restoration, there's a gadget called a wonderbar which allows the brush-battler to apply herbicide to a cut stump without stooping. Another gadget's called a weed wrench, which is a vise levered onto a long handle that allows the worker to prize out and uproot woody undesirables.

Participants in the stewardship network seem to favor old-fashioned tools, such as scythes, hatchets, and drip torches. I noticed in Steve Packard's hard-working car an enigmatic little object, a short, smoothed cylinder of wood with a small circular band of metal attached. I learned that it was a *snath*, the detachable handle of a scythe. Low-tech, and with a romance all its own. Clearly part of the knack of organizing volunteers is entrancing them with the work, by all manner of means.

Laurel Ross's transitive goodness has to do with connection—her connection with the people working as stewards on the prairie and savanna restoration projects dotted throughout Chicago; and the connections she effects between people and the communities of flora whose revival is the network's reason for being. One gets the impression that she, and her colleague Steve Packard, are never not busier than mere mortals could, or should be. Their projects, wildly successful in many respects, are extensive (27,364 acres, 207 sites) and intensive (about one steward per site, 3,809 volunteers, over 50,000 hours worked in 1992). It's also abundantly clear that these people love their work, thrive in it to an uncommon degree. Ross exhibits an appealing combination of sensitivity and practicality, and general wisdom, sources for that latter quality likely being her engagement with its two biggest wellsprings, Nature and Culture. Hope does not seem to be a problem for her. Her stories of restoration endeavors and volunteers are colored by delight. No misanthropy in this kid. In February 1993, downtown in Chicago, Ross spent an hour in conversation with me, speaking of the rewards in the work.

The reason that people are so willing to work so hard in the Volunteer Stewardship Network," says Laurel Ross, "is that they are looking for something meaningful to do. People think of raising their children as important. People think of making art as important.

This is right up there. It's more important to a lot of people than their jobs.

"So many of the life forms we're dealing with are in such serious trouble. Endangered and threatened are two very strong words, and rare's pretty powerful, too. Sometimes I wonder why everyone isn't hysterical about the loss. Of course the reason people are so interested and involved in the stewardship network is that most times when people get an inkling of how important action is, they don't have a clue as to how to do it. The hole in the ozone, for instance: I'm upset about it, but I really don't know what I can do about it besides political things. The Volunteer Stewardship Network offers a way to act, and not just in a tiny role.

"This year we're starting a five-year project as a part of the federal recovery plan for the prairie white-fringed orchid [*Plantanthera leucophaea*], which is threatened. Illinois is mostly where it used to be—we're not on the edge of its range or anything—there should be lots of it. Volunteers will be going out to where there are a few remaining populations and pollinating the flowers, with toothpicks. We'll be coming back later and collecting a portion of those ripe seeds and taking them out to places where that plant should be and once was and planting those seeds and monitoring the plantings year after year. About five years later we may see some flowers—that's about how long it takes."

Ross later added that "one reason the plant is so rare today is that the hawkmoths that pollinate them are not finding the widely scattered populations. The main reason that *they're* so rare is that their habitat has become so degraded. As we improve the habitat through stewardship the orchids will be able to thrive. Pollination guaranteed, seed dispersal enhanced, habitat improved—we're interrupting the cycle that's eliminating this plant; a holistic approach really. We're only restoring the plant to places where they will be protected—through management and through legal protection.

"If you had to pay scientists or even students to do the work, it would be expensive—it's so time-consuming and labor-intensive. Yet it's easy work to do. . . . You could, anyone could, save a species from extinction!

"I like organizing," Ross continued. "It makes me feel that if there's something wrong, there's something you can do about it. My practice involves getting excited and jumping in head-first and then figuring out, once I'm going down for the second time, what parts of the work to keep and what parts to not keep. If I tried to think it through in advance I probably would be a lot more timid.

"A lot of my job is matching up people with jobs when there's all this work to be done, trying to figure out who would be good at what and how to get them started in it so they'll be successful and it will be good for them and good for the project. People really appreciate being tempted by big juicy projects that they might not think up for themselves.

"A lot of times you'll describe a project and people's first instinct is to say, 'Well, I could do two percent of that.' I think they really appreciate being led to understand that they could do sixty percent of that—a hundred and ten percent. It helps them think more about their abilities.

"The experiences get bigger and bigger—you're girdling whole trees, and you're setting fire to the woods, setting fire to the prairie with these big roaring flames—you're actually making choices about the course that the management will take.

"I don't mind it if we make some mistakes. Usually we'll notice it and stop making them. Somebody will learn something from it. You can eventually set it right. Some mistakes are pretty costly but it's a lot riskier to not do anything. We're pretty sure that if we don't do something it's going to be a lot worse."

In a practice of accentuating the positive to eliminate the negative, Ross said, "This year we're trying to manage Bluff Springs Fen [an extraordinarily rich ninety-acre mosaic of prairie, fen, and savanna ecosystems forty miles west of Chicago] as a model preserve. We want to get a lot of people in there but in a way that it won't hurt the preserve and so we're going to organize a whole series of little things. Every weekend there'll be something going on—there might be a Saturday morning bird walk with ten people. There might be a Friday night sedge identification walk. We want people in there because we want people to love the place. We also want people in there who are doing good things we like there, because there are also people doing really bad things in there—people with dirt bikes, people having messy parties, and doing some pretty destructive things. Cops are not the only answer to that, fences are not the answer to that, and laws aren't the answer to that. What we've found is that having people in there—our people in there—is a big part of the answer to that. People don't want to go where there are people who might see them at their mischief. The more people who know what a rare thing is and what their effect on it is, the more people are going to act appropriately.

"We had this little boy last fall along with us as we were seed collecting. He was so taken with it and he was so smart. He came up to me afterward with this one rattlesnake master seed in his hand and asked permission to take it home and plant it. He was so respectful!

"People do develop almost a patriotism to this magnificent midwestern landscape. It's what we have here, and it's available every day. Savanna is not something people grew up longing for, or thinking about, or writing poems about. So this is a change.

"The network is so good at peer teaching. People do take it as a real responsibility to help other people see what they're seeing. That's how I've learned most of the plants I know—by being out there with people who know a different one than I do. John Balaban and I were walking through the Miami Woods Prairie and he bends down and says in a sort of a teacher way, 'Laurel, do you know what this is?' and it's a tiny thing about a quarter-inch high. I bent down and looked at it and it was a baby rattlesnake master. If you looked at it closely enough you could see that that's what it was—it had all the characteristics. That was a great moment. [Rattlesnake master—*Eryngium yuccifolium*—a member of the parsley family, is a distinctive character in the grassland community. It can grow to be five feet tall, which is why John Balaban's ability to recognize it as a tot is virtuoso botany.]

"Last fall we were clearing a huge area and a little boy and his dad were discussing the work—we had been explaining what we were doing and had talked about the shade and the light and so forth, and this kid, as if he'd been scripted, said to his dad, 'Remember how dark it was when we came in? Look how light it is now! Golly, these flowers are really gonna be able to grow now!' And his dad said, 'You've got to have the *photo* for the *synthesis*.' That boy will never forget that—he *learned* it. That's the kind of thing that we're hoping will happen with kids—that the way things grow won't be just some sort of book stuff that doesn't mean anything to them, that they'll actually get it because they saw it, they did it.

"Have you ever seen any of the crazy newsletters that are produced by the Volunteer Stewardship Network? There are dozens! Every region has its own newsletter that goes out to two or three hundred people. There's all this obscure stuff being written on subjects like how to build a brush pile—there are people who clip things out and put them on their refrigerator, I'm sure.[5]

"We talk about the 'restoration community'—it does feel like a community, and sometimes it's more like a cult. After three meetings three nights in a row discussing all this stuff I begin to think I've gone a little beyond normal range. In the height of seed-collecting time I sometimes question how much time I'm spending on it. My personal volunteer role is as seed-collecting coordinator of the North Branch Prairie Project. It has its moments. There have been times when my house has been filled with bags and bags of seeds and little bugs that have come out of the seeds crawling all over.

"Doing something with all of your heart is irresistible. My job is so filled—packed—with delightful people, people who are striving to be productive and happy and do a good thing. You can't help but be positive if that's what your experience is day after day."

To spotlight Laurel Ross, who for years volunteered for the North Branch prairie restorations before she became a full-fledged staff person at the Nature Conservancy's Illinois field office; and Steve Packard, whose intelligence and charisma have stimulated and animated much of the work, is to ask these two people to stand for a great many others. They would certainly demur from taking much of the credit for this remarkable program of earth healing. But it's only human to focus on individuals rather than on populations—the many hundreds of volunteers who bend their backs to the task—so I must ask the reader to pardon the synecdoche.

Packard and Ross both seem averse to ideology. This pragmatism, as matter of fact, is very much the institutional culture of their employer, the Nature Conservancy, and of most land conservancies. You don't raise sums large enough for the purchase and protection of sizable tracts of land by denouncing the failings of private ownership and the excesses of capitalism. Quite possibly Ross and Packard feel that empowering thousands of volunteers to save species in northeastern Illinois, indeed to be encouraging the rescue of savanna ecosystems throughout their former range, is plenty to accomplish; and that the practice of restoration, inherently transforming, is enough.

Chapter 8

Salmon Support

Water flows downhill and salmon swim upstream. Half a century ago in the Mattole River watershed, water made its way down thousands of nameless rills, then scores of creeks, to become the Mattole River and to wind its sixty-two miles to the Pacific. Before the time when nine-tenths of the watershed was logged, the water's movement was governed just enough by the forest of Douglas fir and tan oak and myriad plants of the understory that there was an equilibrium in the Mattole's tributary streams and riverbed. There were shaded creeks, clean spawning gravels, deep pools, and a cool estuary whose brushy banks harbored an abundance of insect life, all of these conditions being necessary to good salmon habitat. So the Mattole, like virtually all the rivers pouring into the Pacific, from Hokkaido to Monterey Bay, supported a good population of wild salmonids—native races of king and silver salmon, and steelhead trout. Descriptions of Pacific Northwest salmon runs beggar the imagination with their abundance, like the tales of flights of passenger pigeons numbering in the hundreds of millions or of bison herds blanketing mile upon mile of the plains.

"Stories have been handed down to us," wrote Janet Morrison, chairwoman of the Mattole Restoration Council, "of streams teeming with salmon, spooking horses at crossings, and of men with pitchforks standing on shore and pitching the salmon into wagons."[1] The situation is different today. In 1991 so few king salmon returned to the river that no salmon eggs could be harvested for the Mattole Watershed Salmon Support Group's hatch-box project. (The hatch boxes, where salmon fry can be reared, are a crucial part of a holistic effort to keep the river's native salmon run from extinction.)[2] In 1992 there were a few adult spawners. Their race may yet be facing the fate of the passenger pigeon.

Whenever we try to pick something out of the universe, said John Muir, we find it hitched to everything else. Some years ago, when these Mattolian back-to-the-landers, or "New Settlers," took it upon themselves, for reasons both mythopoetic and practical, to foster the resurgence of the Mattole's native race of king salmon, they learned they had to follow the salmon's existential logic right up into the root hairs of the watershed's upland forest. A precious remnant of once-great diversity, the Mattole king salmon are among the few scores of wild populations remaining of the thousands of races that had enlivened the tributaries of the North Pacific.

To the uninitiated a salmon is just a salmon: lox, maybe. To the aficionado, the five species of Pacific salmon spawning in North America—king, silver, red, humpbacked, and dog—have distinct identities and seasons and flavors, all arising from their home streams. Like characters in Russian novels, their names vary from situation to situation, and from phase to phase. Those same five species also are known as chinook, coho, sockeye, pink, and chum salmon. From hatching through maturity they are called alevin, then fry, then parr, then smolt, and finally adults—spawners. All that language bespeaks the character and complexity of these anadromous fish which run back to the river from the ocean to mate and bind together the land and the sea in the course of their existence.

The salmon was to the first peoples of the Pacific Northwest what the bison was to the peoples of the prairies. Salmon were a staple of existence, but no mere meat. Among the Indians who caught them, a ceremonial relationship to the salmon was virtually universal. They were regarded as conscious, immortal beings. Throughout the region, the taking of the first salmon was a ritual act, with prescribed formulas of address: "I am glad I caught you. You will bring many salmon into the river. Rich people and poor people will be happy. And you will bring it about that on the land there will be everything growing that there is to eat . . ." (Yurok First Salmon Ceremony).[3]

Taboos dictated which persons might partake of the first salmon, and how the offal was to be disposed of, especially the heart—that there should be no mutilation of it by scavengers. The purpose of these courtesies was to show respect to the salmon and to ensure its return from the sea and thus the people's well-being throughout the year. The salmon were addressed by many different names: "Chief Spring Salmon," "Quartz Nose," "Two Gills on Back," "Lightning Following One Another," and "Three Jumps."[4]

It is not so difficult to understand why these totem fish (and of them, the king is the biggest, and the mightiest traveler) enjoyed such great respect. Their lifeway is an heroic evolutionary saga of wild Nature and place. The great silver bodies hurtling headlong up rushing torrents to mate; the careful excavation, in her natal stream's gravels, by the spawning female making her redd, or nest; the arching, simultaneous expression, by female and male, of eggs and milt into the redd's pockets; and the wasting and death of the spawned-out fish whose corpses feast the bears and bald eagles streamside—it all is nothing short of Wagnerian.

Unlike Atlantic salmon and steelhead trout, Pacific salmon species mate once only and then die. They hatch and grow to a degree of maturity upstream from the sea. The length of time they spend in the freshwater pools and riffles of their origin varies by species. At some point, though, a change in the endocrine system creates an absolute need for salt, and the young salmon, now called smolt, commit themselves to the trip downstream to the ocean.

There, again depending on the species, they spend one to five years eating, trying not to be eaten, and growing from six inches to a yard or more long and accumulating from six to sixty pounds of rich flesh. Their movements in the Pacific are presumed to be great northwesterly gyres. The distance they venture off the continental shelf is a matter for speculation. As the salmon approach sexual maturity their bodies prompt the return to the home stream. By means largely olfactory and perhaps geomagnetic, they navigate the cloud-locked Pacific, enter rivers as vast and contaminated as the Columbia, Sacramento, and Willamette, and, where the way is not barred by a dam, return to a home stream, one watercourse among thousands, to reproduce.

This inexorable longing for and loyalty to a natal spawning stream that can call across a thousand miles of stormy sea means that the various races of salmon that return to all these rivers and creeks have over time—a long time—become genetically distinct. Of the thousands of native races of Pacific salmon, a few hundred remain. Recently the American Fisheries Society has declared 124 races of Pacific salmon to be in some degree of danger of extinction. Among these is the Mattole king, which is at "high risk" of perishing.

Epic traits of the Pacific salmon's life cycle—the glinting infancy in streams and maturation at sea, the dramatic return, full-grown, to the same spawning grounds they hatched in—are also what make the salmon almost a self-catching fish. It's an obvious trick to spread a net or weir across the stream and haul them in. Because the various salmon runs occur at about the same time each year and because the fish arrive *en masse*, an entrepreneur can just about catch himself an entire generation of fish. After all the spawners are taken over several consecutive years (five years being old for a salmon and about the longest time a stream's cohort of fish might be at sea), that stream's race is extinct. This practice of trapping a stream's whole run annihilated numerous races of salmon before the turn of the nineteenth century, when it was outlawed. It comes under the general heading of "overfishing," which by other means continues to this day. Now the netting is done offshore. High seas driftnets—miles-long webs hovering in the open ocean, indiscriminately capturing all manner of marine life that comes their way before being hauled aboard factory ships—are being eyed as a culprit in the radical decimation of the salmon, and a great many other marine creatures as well.

The development of the American West has meant the deconstruction of the salmon's world. Dam building, in this century, especially the construction of high dams, decimated salmon runs throughout Idaho, Oregon, Washington, and Northern California by flooding river valleys and drowning their spawning streams or simply by rendering the streams inaccessible. Fish ladders were supposed to enable spawning salmon to continue upstream past these great concrete obstacles, but they have not been very useful. Most grotesque is the mangling of salmon that are sucked into the turbines of power dams. During the California gold rush, hydraulic mining swept whole landscapes worth of sediment into stream channels, destroying fish habitat by clogging the gravels that harbor eggs, alevin, and fry, and leaving behind a lunar landscape. In many once-forested places, among them the Mattole, converting forests to pastures, clear-cutting and the road building that goes with it (as well as with every new homestead), and poorly managed grazing all have degraded lands upslope and salmon habitat downstream. Thus it develops that if you want to restore the fish you have to heal a whole watershed: vegetation, erosion, social fragmentation—the works. For the better part of this century, the palliative response to the widespread destruction of salmon spawning grounds has been to stock rivers and creeks with generic, hatchery-bred salmon fingerlings. In his essay "Salmon of the Heart," writer Tom Jay says fishermen refer to these hatchery products as "rag." He himself calls them "homeless seagoing spam."[5]

In order to get next to some of these charismatic fish, albeit the spam version, I drove twenty miles from my home over to the seventeen-year-old Platte River Anadromous State Fish Hatchery, largest in Michigan, and a source of some of the millions of Pacific salmon fingerlings that are planted in rivers throughout the state. In the mid-1960s the salmon fishery was created to use the exotic, carnivorous salmon to check the plagues of alien fish species in Lake Michigan, particularly alewives. Later the sporting possibilities of salmon fishing were grasped, and the outdoor recreation dollars fishing licenses bring to the states bordering the lake argue for its continuation. Whatever their provenance, salmon are big and exciting to catch. (They're also fatty enough that scientist James Ludwig mordantly refers to them as a clever means of capturing the toxic fat-soluble polychlorinated biphenyls [PCBs] that contaminate Lake Michigan's waters. People should be very cautious about eating Great Lakes fish, he allowed. "Anyone who wants to reproduce had better be damn careful," was the way he put it.)[6] Despite their occult toxicity, these coho salmon provide angling pleasure to fishermen and women to whom the idea of native species is no big thing (and who, given the upheavals of the aquatic fauna in the Great Lakes, would be flat out of luck if it were).

The hatchery was an Orwellian dystopia for fish—a big concrete industrial site, with elaborate hydraulics and pneumatic feeding systems and a fancy Italian-made salmon egg sorter. In the hatchery, things are tilted toward industrial efficiency; toward a high yield of a reasonable facsimile. The adaptations of hatchery fish are in the direction of tolerating crowding. They are bred for fast growth on pelletized food, and then for being adaptable to the live foods they'll find after stocking so that they will get bigger yet. They are virtually domesticated, deracinated, infantilized, denatured (the analogy with humans in mass society is uncomfortably clear). The worker who gave me the tour, though, really seemed to have some respect for the ethos of the salmon, an understanding of the forces that had shaped its character, and a knowledge of the conditions that these still-remarkable fish need after they are pumped out of the trucks that transport them to the streams where they are planted. From there they will enter the lake and start traveling for a few years south to the shores of Indiana, then west to Wisconsin, and finally back across Lake Michigan to their foster streams, dead-heading sixty miles east in three days.

Ever since the sea lamprey snuck into the lower Great Lakes through the Welland Canal ("The lampreys came in right after the lakes had been connected by the St. Lawrence Seaway to the Atlantic," comments David Simpson, a Mattole salmon supporter who grew up fishing on Lake Superior. "What a great parable of the subjugation of

an enormous, complex ecosystem to the brutishly simplistic requirements of World Trade."), the Great Lakes' ecosystems have been like
a fishbowl belonging to some kid who can't quite get the combination
right. The sea lamprey, whose parasitism of the native lake trout eliminated the Great Lakes' top carnivore, have wrought havoc but are not
the sole culprit in the disequilibrium. As with the Mattole, environmental degradation from logging played a part in extirpating the original lake trout populations. (Introduced lake trout are beginning to
establish themselves now, however.) Overfishing was another factor
that early in this century upset existing predator–prey relationships,
and resulted in situations conducive to booms of alien fish like smelt
and alewives. By the 1960s they were using front-end loaders to clear
beaches in Chicago of alewives. These several aquatic extravaganzas
have taken place over the last half-century or so, and have sent fisheries
managers scurrying in attempts to arrive at a functioning species mix
that will serve human, and rough biological purposes, if not restore
the ecosystem. Thus the hatchery system, whose primary purpose is
to supply a predator to Lake Michigan's crazy, mixed-up aquatic
fauna.

In contrast to the clony ignobility of hatchery fish, wild salmon
embodies fealty to a particular watershed, serves to define that living
place, and, inseparable from it, is shaped by it in turn, much as a red
blood cell is part of a larger being, interdependent with millions of
other cells and kinds of cells, constituent particles of limbs and systems. The analogy, of course, doesn't carry all the way, but the sense
of the watershed as an organism with its own juices, organs, members,
and health does suggest how to begin to convey a feeling of the sin-
gular identity of a life-place. And how a group of twentieth-century
reinhabitors, one by one, might come to love their watershed and its
most charismatic species; how they might find their lives bound up
with its. In the Mattole, the prime value of the salmon is that they're
wild and indigenous. The Mattole Restoration Council (MRC) respects the venerable wisdom of the fish's co-evolution with the watershed, and in its restoration work the council's members strive for a
high degree of fidelity to nature.

If the Mattole Restoration Council, "an organization based on watershed priorities to serve the interests of the various factions of the human community engaging in recovery," and the Mattole Watershed Salmon Support Group (MWSSG) didn't exist, bioregionalists would have had to invent them. These groups perfectly embody the reinhabitory concept of "becoming native again to place." I had been hearing stories about the MRC's work for a decade, years before people began talking about ecological or environmental restoration. The stories suggested a sensitive and strenuous remarriage of humans and entire ecosystems, a way of living-in-place that aims at redressing some of humanity's juvenile errors of land exploitation, and possibly a way of seeding a new culture out of those amends. Salmon support and watershed restoration as practiced here constitute a long-haul intention of working not merely in historic but in geologic terms.

When it began in the 1970s, the Mattole Watershed Salmon Support Group, which was a founding member of the Mattole Restoration Council, just wanted to enhance the reproductive success of the king salmon. First they made hatch boxes, simple nurseries where eggs taken from and fertilized by wild salmon trapped in the river could be incubated. In the clean filtered water flowing through the silt-free gravels in the hatch boxes and rearing pond the fry could grow, enjoying regular meals and an absence of predators, and then be released into their parent creeks.

"To enter the river and attempt to bring this strong creature out of its own medium alive and uninjured is an opportunity to experience a momentary parity between human and salmon, mediated by slippery rocks and swift currents. Vivid experiences between species can put a crack in the resilient veneer of the perception of human dominance over other creatures," writes Freeman House. "Information then begins to flow in both directions and we gain the ability to learn: from the salmon, from the landscape itself."[8]

What the salmon supporters learned was that erosion problems upstream had radically changed the seaward reaches of the Mattole River, widening its bed, stripping away shade-providing vegetation on the banks, and filling in the pools where the salmon, which love cool depths, could come to oceangoing maturity. Torrents carried away nesting gravels from creeks and silt washed down and smothered what was left. So to save the world—of the salmon, at least—the Mattole Restoration Council members began over the years to plant thousands of Douglas fir seedlings and native grasses on cut-over lands upstream; they got into "restoration geology," sometimes using heavy equipment to put structure in streams which had been rendered too simple; they sweated to armor eroding creek banks with head-sized cobbles; to transplant bits of native vegetation that could stabilize sections of eroded stream banks; to plant willow cuttings and alder seedlings alongside the river to shade the water and foster the insect life on which the fish depend.

In the Mattole Watershed Salmon Support Group's scrapbooks are snapshots of ten-gallon buckets of salmon fingerlings being packed on horseback down to streams where they'd be released. (Horseback. In places where hatchery-bred salmon are stocked in streams, the fish are transported in specially cooled and oxygenated tank trucks.) In other years, with the help of the young people in the California Conservation Corps (about whom they had much good to say), they've winched boulders around in the creeks. Most everything they do is hard work: moving rocks and breaking up log jams in these deep, steep creek bottoms. The people laboring to help the Mattole heal are so tough and committed, they're acting as though restoration was their only choice. The erosion problems they're contending with in their life-place are formidable, but no more so than their will to remedy them.

For instance, there's a landslide in the Mattole watershed that is a mile long from ridgetop to river, and a half-mile wide, 375 acres of slump or debris flow that cut loose during an intense spring storm in 1983, the rains slicing down and ruthlessly cutting into exposed, unsteady soils. That monumental Honeydew Slide didn't happen overnight, but was a consequence of many different instances of inattention, from clear-cutting to careless road building to a failure to deal with small erosion problems before they grew large. More amazing than the magnitude of the slide was the reaction of the doughty Mattole Watershed Salmon Support Group, recently formed and ready to rush to the rescue in the event that the Honeydew Slide blocked the river and cut off some of the best and most heavily visited salmon spawning and rearing areas in the whole river basin.

According to the Mattole Restoration Council's newsletter, the MWSSG "had formed emergency contingency plans with the [California] Department of Fish and Game to actually move the winter run of salmon upstream if they got stranded below the debris dam. . . . The group was ready to lug the piscine paramours upstream in buckets if necessary." (These extraordinary services proved not to be necessary to the salmon, for, the newsletter continues, "with the rains, the river rose and cut a new channel south of the debris dam, providing fish passage.")[9]

Work in the Mattole is about preserving the spark of diversity that its native races of salmon possess. It's endangered species preservation work in the real world, a place where much of the land is privately owned and can't be acquired to create refuges that may be managed at the pleasure of biologists. Members of the MRC and other such groups in the valley see restoration as an indigenous occupation, like tree cutting, farming, fishing, or ranching, a livelihood peculiar to the locale, one that might continue for generations. This is a different sensibility than hiring restoration done, or imposing a restoration design from without. It is what can be done when it's too late to conserve. The object is not to reinstate some static idyll, but to restore the dynamic of evolution in an ecosystem and to include the human in that dynamic. "It is comforting to envision what benefits might result from watershed-wide restoration work (and wise maintenance of public lands)," wrote Freeman House and David Simpson in 1983 in the inaugural issue of the *Mattole Restoration Newsletter*.

If you can, imagine starting from the ridgetops and headwaters . . . planting trees and grasses for slope stability and future timber . . . as roads get built and maintained so that erosion slows rather than increases. The river gradually flushes itself and stabilizes. Vegetation begins to seal it in a cooling shade again. Work in salmon enhancement begins to pay in visible increases in spawning runs. Silt washed off the upland slopes begins to deposit itself permanently in rich alluvial flats. Grains and vegetables grow in soil that was formerly swept to sea.

A generation from now, our children reap a harvest not only of fine timber, abundant fish, productive grasslands and rich and varied plant and animal communities—but also a tradition which will assure the same harvest for their children.[10]

Perhaps it's possible to place restoration activities along a spectrum, and also to propose some differences between restoration and reinhabitation—not an absolute distinction, certainly, but one as real as that between restoration and rehabilitation. Restoration implies an exacting fidelity to the original; rehabilitation may resort to the use of similar species in order to create a rough, but functional, semblance of the original ecosystem. Restoration presently, and in many cases necessarily, requires that access to the recovering ecosystem be restricted, rather like the burn ward in the hospital. Reinhabitation implies living in, and having an economic stake in, the place restored, not in the touristic sense of being able to charge admission at the park gate, but in being able to derive what House calls "natural provision" from one's own ground: free (but not easy) protein, fuel, and building material. Restoration does not pose alternatives to the socioeconomic system that is necessitating restoration; reinhabitation does. It means the articulation of a culture of place, and for this to come about in our time means the restoration of human community in a society whose members have wildly differing and fiercely held ideas about what land is for.

Jan Morrison, the chairwoman of the Mattole Restoration Council, is a land restorationist. In her last year as a communications student at Humboldt State University, Morrison headed for the hills of Redwood National Park to be part of a small business engaging in erosion control projects. It got her outside, planting trees, and provided good comradery.

When we met in the Mattole Valley in the late winter of 1990, Morrison spoke about her love of restoration work, the pleasure she took in making willow wattles, bundles of willow withes to be placed in horizontal trenches with tipped-back berms cut along eroding slopes. The bundles themselves help to check the erosion, and the willows sprout, as is their wont. Morrison described hillsides that had been mended in this way as looking as though they'd been embroidered with the lines of emerging green shoots.

"I learned just by doing," she says. When the erosion control group broke up in the early 1980s, she moved to the Mattole to work for a nonprofit group that had a contract from the Fish and Game Department to do a watershed-wide survey of all streams where the salmon were spawning. "They'll pay you for walking in the streams," is what she'd heard. The prospect appealed to her. The work led to her increasing involvement with the Mattole Watershed Salmon Support Group. "Salmon people were always in the river," she said. Immersion in the watershed—field study of the fish, careful evaluation of the usefulness of different erosion-control techniques, creation of a home-grown information base—is essential in this reinhabitory work. A lot of what the salmon enhancers do is mapping. It's an essential requirement for providing a local reality check on the generalizations issuing from the larger agencies that concern themselves with land and water use. The indigenes see no evil there, simply a lack of fine-grained knowledge. Morrison spoke about all the microclimates within the watershed, for instance, and said that residents of various drainages keep their own long-term rain-gauge records of the Mattole's monsoon precipitation. If information that detailed were incorporated into the regional weather maps, Morrison said, "it would tend to make for untidy isobars." It is this very untidiness that ordains the real-life conditions in which land healing may go on. There's a lot of passionate knowledge of place among the salmon people. They're willing to do the long-range and baseline studies in order to be able to make effective comparisons and to account for what they are doing. Not only do they have to wade out into rushing streams when the salmon are running, and wrestle with these great beasts, but they also have to keep accurate records of how the many different elements of the watershed restoration are working.

"Just trying to save a watershed is a lot of work," says Morrison. It's not only physical work, although there's plenty of that. It also involves jurisdictional diplomacy. The California Department of Fish and Game basically governs the fate of the salmon and it must sanction rearing activities. Even picking up fallen logs across a stream requires a permit. The Mattole Restoration Council was formed to provide an aegis for the numerous groups throughout the watershed, for the neighborhood associations and creek councils that wished to play a part in the salmon work but lacked suitable status. More important, commented Freeman House, "decisions that affected a whole bunch of things were being made by too few people. The MRC was formed to provide a broader inhabitory decision-making base."

Several months after my visit to the valley, talking on the telephone with Janet Morrison, who had just come from one of the grueling community meetings that seem to be a staple for salmon enhancers, I began to get a sense of the tact necessary to these counterculturalists, and to admire their rhetorical self-restraint. Another thing that became clear is the considerable degree of organization in the Mattole. For about 3,000 people in a 300-square-mile territory, there are surprisingly many groups and associations, all with real claims to a place at the council fire. It makes Chicago's Cook County seem almost random.

In the Mattole, there are really two separate cultures, the rancher and the homesteader, living side by side, Morrison observed. As in so many rural areas, there's a historical as well as an economic stratification of residents. The first peoples were extirpated without a trace. Then came ranchers, then loggers, then hippies, then dope-growing hippies, then second-home developers, then all manner of paranoias and antagonisms to rive the community through and through. There are lots of different stripes of love-hate relationships with the government, be it federal, state, or local.

A recent government action precipitated the beginning of a watershed-wide process for developing consensus on land-use practices in the valley, and ventilated some of these passions in the process. A State Fish and Game Department memorandum on degradation of salmon habitat had urged zero net sedimentation. This meant that loggers must either prevent sedimentation as a result of their activity or mitigate erosion damage elsewhere. The memorandum became the charge of the State Department of Forestry.

To help acquaint valley residents with the new policies, the Department of Forestry called together a public roundtable. At its second meeting, said Morrison, a good turnout degenerated into a shouting match among the ranchers and homesteaders and environmentalists. "Let's go outside and settle this" was heard. Somehow an almost too-good-to-be-true outbreak of peace occurred. Someone suggested an agenda and another meeting. A rancher and an environmentalist stood up hand in hand. It began to resemble an encounter group, she said. Dan Weaver, a retired Navy pilot skilled in conflict resolution, offered to facilitate subsequent meetings. Attendance increased from a dozen to thirty, forty, then fifty people. Increasingly, they are working together.

Salmon enhancers are trying hard to develop a cultural sense of place across the generations. Every year or so the valley kids paint a mural in their school. The first time they chose salmon in creeks as their theme, said Morrison, they portrayed the salmon going every which way. Other details of the mural's natural history were inaccurate as well. Over the years since the hatch-box program has been working, the valley's schoolchildren have been invited to participate in this release, trekking down to the streamside and dumping big buckets of little fishes into the flowing waters of their destiny. It's part of reinstating salmon in local culture. After the children had participated in the salmon release, they painted another mural. In this one the salmon were all headed in the right direction—upstream.

The Mattole River watershed lies about 200 miles north of San Francisco, then a winding highway west maybe fifty miles through a redwood forest and over a couple of ridges of the California coast range. The place is plenty remote and, by me, primitive. Dank with coastal fogs, the Mattole is said to be the wettest place in California. The country is so steep and broken that it claws the water right out of the air and pours it down a million little creeks. The landscape is overshadowed by ridge and range, cloaked in olive-drab prairie, Douglas fir, and patches of brush. In the late winter of 1990, I made a visit to the Mattole. There I enjoyed the hospitality of Freeman House and Nina Blasenheim, who have lived and worked in the valley since 1980.

House and Blasenheim and their daughter Laurel live very simply, and this, too, is part of reinhabitation: a voluntary simplicity that asks as little of the Earth as possible. They heat with wood, which appears to be plentiful. California's climate is forgiving enough that they can get away without insulating their great big barn of a house. They have an open-fronted outhouse and an outdoor shower. Their household utilizes propane, which seems really to be the salvation of the rural homesteader, running their gas refrigerator, gas lamps, gas cooking stove, and a demand gas water heater—they do have all the hot water they can afford the gas to heat. A photovoltaic cell provides the juice for additional, electric illumination. They communicate by radiotelephone. Nina Blasenheim was vegetable gardening that late winter I visited, and the chickens were roaming free through the rows.

In addition to being one of the wettest places in California, the Mattole is also probably one of the most seismic places on the planet. Three tectonic plates come together in a place offshore called the Triple Junction. They had a pretty fierce earthquake shortly before my visit in 1990. Blasenheim talked about what it felt like to be at the epicenter rather than somewhere out where the wavelength is longer. She said it was as though the earth had been playing crack-the-whip. In the course of a day's visiting, touring, and hiking, House showed me some evidence of earthquake damage in the valley: a little fissure in the earth up near the road; slides, lots of slides—saturated soil slumping down; root bundles coming undone and dragging big clumps of dirt and trees down to block the dirt roads clinging to the hillsides. It's turbulent earth. In April 1992, three earthquakes, ranging from 6.0 to 7.1 on the Richter scale, struck the Mattole, causing a meter's worth of orogeny in half a minute, changing the intertidal zone as well as life onshore, where about half the homes in the area were rendered uninhabitable. Yet this violent unpredictability of the terrain is regarded, and respected, as an aspect of character, rather like the short temper of a generous friend.

On our tour of sites where salmon restoration work was under way, House took me to a spot on Mill Creek. The MWSSG had done some work there to restore silver salmon habitat. It was surely one of the loveliest places I've ever seen, a poetically beautiful dell. It's California-steep—not quite a ravine, overhung by gnarled, leggy tan oaks, these brilliant with lime-green mosses. The canyon is lush with ferns of several kinds, sword ferns most notably, whose root systems House credits with holding up the creek banks. There is also another small plant lovable for its rhizomes. A transplanted clump of *Whipplea modesta*, said House, can help shore up an eroding patch. Fallen Douglas firs catch and bridge the glen, musical with the sound of falling water.

Logging-caused erosion had scoured the creek and steepened its fall such that its waters flowed too swiftly to drop out, or recruit, the right-sized nesting gravel for salmon redds. The structural solution was to winch a windfall log down from the cut-over north bank. The restorers hauled in a portable saw mill to quarter the log lengthwise. They placed the four pieces across the stream to catch sediment and slow it down, so the right-sized gravel fell out. The water falling over the straight edges of the dams scours out congenial pools beneath. House speculated that the six or eight people who worked 100 hours on Mill Creek may have created habitat for perhaps thirty mating pairs of silver salmon. It seems to them a fair exchange.

Part of the value of restoration work is firsthand teaching of the laboriousness of an entropy battle—how very difficult it is to put a casually torn-apart ecosystem back together! Perhaps if more people could feel the ache of that difficulty in their bones, there'd be less tearing apart. That enchanting spot in Mill Creek was human-altered. First, for the worse, by the frenzied logging upstream. But the healing—changing the stream gradient with those Douglas fir dams—left evidence of human activity, too. Other evidence is the rock armoring, dry masonry artfully shoring up the creek banks. Now there are deliberate patterns, straight lines, right angles, planes of logs over which cascade linear waterfalls. The very music and echo of the creek has been tuned by human activity. Mill Creek is not a wilderness preserve; human artifice there aims, ultimately, at restoring natural provision so that the salmon will feed the people and the forest will supply immediate human needs again. It's bold thinking that humans could participate in the ecosystem in a benevolent, postmodern way. The "conscious gamble" of those working in the Mattole is that the interaction with a watershed could engender a moral check on the human impulse to control and determine, expand and exploit. It was in talking with House that I first began, however reluctantly, to question that convenient fictive absolute of a hands-off policy toward wild Nature. To breach that concept means relinquishing *Homo sapiens'* guilt and self-loathing over what human involvement in the landscape has resulted in thus far. And that passage will be sustained by maturing beyond remorse into reinhabitation.

Now the headwaters, says David Simpson, a founder of the Mattole Watershed Salmon Support Group, is in healing mode. As of 1993, California's drought was declared to be over, which means that one of the multiple threats to the salmon's survival is in abeyance. The threat this natural disturbance posed to the radically diminished salmon population might have been less mortal if other conditions also hadn't been so adverse. The problem is that by now we've driven so many species into tiny little corners that the natural disturbances that once developed their evolutionary strength of character no longer play a salutary role, but can instead threaten their survival.

A reinhabitory organizer might envy the Mattole people, and others in their region, the charisma of the species that is the *genius loci* of their restoration work. One might envy it all—sharing a common purpose in a beautiful surround—save for the heartbreak of facing the possibility of the creature's demise. Getting your destiny bound up with that of another species is a supremely risky form of romance. Yet the constancy of the Mattole restorers has been like that of their totem species; their endeavor to date has always been against the current of land abuse, and their loyalty to place as obdurate as the king salmon's. After a decade of dealing with the minute particulars of rearing salmon fry, of creating plunge pools and elegantly shoring up creek banks and planting thousands upon thousands of trees, the fate of the salmon remains in doubt. Yet the seed bank is there. The possibility is alive, incarnate in the sea, and a small nucleus of community is still bound together and informed by it.

Basically, said Simpson, in a phone conversation in autumn of 1992, "The question of salmon survival is by no means decided, but we're doing everything we can.

"Politically it's a whole lot better; ecologically it's unresolved," he said, and spoke of an increasing coalescence in the views of the people living there—ranchers, homesteaders, even small timber companies with whom they are enjoying "new levels of cooperation"—as a result of their meeting together as members of the watershed alliance. "All we're doing is trying to hang on to what we've got while the habitat restores itself. It's the only hope of recovery."

Simpson likened the nature of this change to the long, slow turn of a battleship. "All of human history is directed toward ecological disrepair," he said, so it is hardly surprising, although no more bearable, that redirecting our lifeway toward healing should be starting when it's nearly too late. "An enormous price has been paid for this consciousness."[11]

Chapter 9

Greenwork in Utopia

It may be that the old real estate agent's truism that "they're not making land any more" will undergo a little revision in Auroville, and anywhere that this devotional, reinhabitory brand of restoration is under way. Auroville is a twenty-five-year-old, 2,500-acre intentional community in Tamil Nadu State in southern India. It aspires to become "a city that the earth needs." When I tell people about Auroville, I find myself always beginning with a description of the ground the Aurovilians, who mostly came from abroad, first confronted. To call it soil would be a misnomer. It's laterite, baked terracotta solid by the tropical sun, although it once was permeated by rich scrub jungle. The Auroville lands are not contiguous, but a patchwork of plots interspersed with temple grounds and Tamil villages, whose inhabitants are, for the most part, very poor. The land that Auroville occupies is the land that the farmers were willing to part with, land that had been deforested long ago then cropped or grazed beyond its limits. A certain social and economic, as well as ecological, depauperization forms the backdrop to the Auroville experiment while the phenomenal spiritual richness of India pervades and animates it all.

Even now, with much of the land in some stage of regeneration, and attractive households, administrative centers, and appropriate technologies the norm, it isn't all that difficult to picture the desolation that was Auroville in the beginning—broad stretches of gullied, parched red earth and not much else. One longtime Aurovilian wrote that in 1969 the land looked like "a once-living earth dying back into a moon."[1] Auroville was where I got a good look at both desertification and large-scale reafforestation. Now, flecking the gullied landscape, one sees the exotic acacias that serve as the beginnings of the new forest. In some places, grasses have actually begun to take hold. And looming in the distances are lines of tall, limber palms, their typecast tropical silhouettes accenting the horizons. Although forest restoration in Auroville doesn't aim at historic authenticity save in a few scrub jungle sanctuaries, it's not on the plantation model, either. In Auroville, reafforestation and all the effort that supports it, effort oriented to a certain kind of human future, a future of living in the land, is called Greenwork. No doubt Auroville got some special impetus from a combination of the drop-out-and-move-to-the-country courage with the go-to-India-and-get-a-guru aspirations that stirred certain members of my generation as the sixties progressed. "What is Auroville?" began a letter from Eric (who, Auroville-style, went surnameless) to the brand-new *Coevolution Quarterly* in September 1973.

It is a new inter-intra national city being built in the middle of a desert in South India near the old French territory of Pondicherry. It has been sanctioned by UNESCO when it was founded in 1968 but it is best explained as the vision of Sri Aurobindo and the Mother: a new city free from international rivalries and internal politics, the site of material and spiritual researches into an actual embodiment of human unity, a place for man to seriously attempt the beginning of the reorganization of the Earth and mankind in a rational and logical manner. It is a commune which escapes the limitations of exclusiveness. The mantram of Auroville is TRUTH, AT THE SERVICE OF TRUTH. The yoga of Sri Aurobindo and the Mother is unique in that it is not religion, no priests, no rituals. It is not the escape from this world into nirvanic states, it is here and now—do it while you can. It is good in the yoga to organize a little of the physical matter around you. The significance of this physical existence in this lifetime is that this is the only field where change is possible, progression, evolution, a new man(kind).

Our work is agriculture, afforestation, village handicraft, printing, handmade paper, . . . geodesic domes, ferro-cement, . . . solar energy, metal working, windmills and borewells, soil rehabilitation and erosion prevention, health and medical care in rural villages, fishing and . . . crystalline structures and moghul tombs, children and education, computers and systems analysis, tetrahedrons and vector equilibriums, running, laughing, playing, and building a new city from scratch.[2]

Part of the significance of the land restoration work at Auroville is that it stems from a discipline, an esoteric understanding of the nature of life and the role of the human being in life. Some say that at least insofar as its utopianism is concerned it is fortuitous that Auroville is in India, that it could be anywhere on Earth. But it is perhaps no accident that Sri Aurobindo Ghose, the sage whose philosophy gave rise to Auroville, was born Indian. Classically educated in England, Sri Aurobindo returned to India and was transformed, through the experiences of revolutionary Indian nationalism and consequent political imprisonment, into a philosopher, and, his followers believe, an enlightened being. Sri Aurobindo would, in 1910, settle in Pondicherry to propound his "integral yoga." The sum of his teaching is not a world-renouncing quietism, but an evolutionary engagement in this world.

Sri Aurobindo's evolutionary philosophy coincided with the experience of Mirra Alfassa, the Frenchwoman who in 1920 would come to India to share in Aurobindo's work. Described as one "born to break the limits," this remarkable woman conceived of Auroville and guided it, as well as the Sri Aurobindo Ashram, until her death in 1973. "If we want to find a true solution to the confusion, the chaos, and the misery of the world," said the Mother, "we have to find it in the world itself. In fact it is to be found only there: it exists, latent, one has to bring it out. It is neither mystical nor imaginary, but altogether concrete, furnished by Nature herself, if we know how to observe her. For Nature's is an ascending movement . . ."[3]

Although Auroville is by no means an ashram (the Sri Aurobindo Ashram is in Pondicherry, and although I can't vouch for the spiritual attainments of Sri Aurobindo's followers, I must say that the entrepreneurial spirit, evidenced by the numerous small businesses in Pondicherry whose names bore the Auro prefix, was lively enough), the few Aurovilians whose homes I entered were, judging by the contents of their bookshelves and their devotional objects, followers of the teachings of Sri Aurobindo and the Mother.

Auroville is, at present, somewhat of a colony, with members of twenty-six nationalities, mostly Europeans and Americans. Of the 900 Aurovilians, about 250 are Indians. It is very much a hybrid situation, West come to the East. Not what you could call Gandhian in the sense of being either nativist or minimalist. Although the original idea in Auroville, which began with leadership from abroad, exotic to Tamil Nadu, even to India, was to develop a community that would be self-sustaining and egalitarian, it isn't there yet.

Auroville was begun in 1968 with grants from UNESCO and other world philanthropies; Aurovilians depend a lot on monies from home, or that they return home to earn. To secure a living beyond the extremely modest 1,200–2,000 rupees per month subsistence that the community provides its full-fledged members, Aurovilians may commute back to their mother countries in Europe, or to the United States to make a little money however they can—pumping gas was a perhaps apocryphal example given by Greenworker Ed Giordano. Aurovilians may be said to come from the ranks of the world's privileged. They live where they do by choice; they have the luxury of being able to subscribe to the belief that, as Giordano puts it, "pushing Nature to get the maximum for our own existence cannot continue." Yet if it were not for the Aurovilians' relative affluence, their experiment might never have been initiated. Although it is by no means the only such intentional sustainable community in India (without even trying I learned of two others, so there must be more), its condition of being a multinational settlement may be unique.

Throughout Auroville are windmills, biogas plants, and solar panels, along with a bunch of other household-level appropriate technology experiments, like Tency Baetens's lagoon pond system for treating his home's black and gray water, or Ed Giordano's house, a sleek, modern cottage solar- and biogas-powered and surrounded by a pleasant shallow gutter which is both aesthetic and a bar to invasions by crawling insects. Giordano's home is off the grid (which in India is mostly unreliable). "Almost all the energy we use here, we produce here," he said. Thanks to a solar photovoltaic system made of components almost entirely of Indian manufacture, "we've never had a power crash."

Auroville's architectural experiments run the gamut, from Auroville's omphalos, the Matrimandir—a dramatic, as yet uncompleted, oblate dome that houses a meditation space and serves as a latter-day cathedral—to airy, thatch-roofed pavilions, to low-slung comfortable bungalows made of cement-stabilized earth blocks. The reception center and guest house where I stayed, the Greenwork headquarters, and the Center for Scientific Research all were attractive, inventive, and simple architecturally.

Auroville has developed and now markets channeled ferro-cement roof elements, latrine cabins, and biogas plants, each of which represents an improvement over traditional ways of meeting those needs for shelter, sanitation, and fuel. However, as Tency Baetens, who's in charge of the ferro-cement works, commented, this material, although certainly more frugal than poured concrete (or wood, at this point), is not, strictly speaking, sustainable, because cement and steel mesh are technologically advanced industrial products whose manufacture is energy-intensive.

I happened to arrive at Auroville the day an "Awareness Workshop for a Sustainable Future," sponsored by Auroville's Center for Scientific Research, was beginning. The workshop attracted a baker's dozen of participants from village development and community organization groups around India, people who were working in projects in organic farming, appropriate technology, and watershed management.

One of the sustainable futures workshop sessions was held at Forecomers, which was among the first settlements in Auroville, and one of the main centers of the Greenwork. Afforestation began as a matter of stark necessity shortly after the first settlers arrived, the location for the settlement being decreed by proximity to Pondicherry and the availability of land. The chore of making the location habitable, therefore, came with the territory. By now there was a goodly amount of vegetation flourishing. That day in 1992 our seminar took place under an expansive raintree.

"What you see here," said Ed Giordano, indicating the vegetation surrounding us, "basically has been placed here."

One entirely understandable impetus to tree planting at Auroville, I instantly grasped, was the need to provide some shade. Although by the time I visited in September the summer's heat was past, it was hotter than I have ever imagined being—and I spent my first sixteen summers in Phoenix, Arizona. Without tree cover, existence in those lands would be nearly intolerable. In the tropics, however, if things can grow at all, they'll grow rapidly. In those latitudes life really wants to live. So trees planted and tended by the first Greenworkers had, in a quarter-century, become almost majestic.

Tineke Smits, chatelaine of the Centerfield Guest House, where I was housed, was among the "Forecomers." As we strolled around the compound during a workshop session being conducted there, Smits recalled in an earlier time walking this area with a map and discovering that parts of the land depicted thereon no longer existed. They had caved away. She and another woman lived in a hut, the two of them spending their first few years in Auroville planting trees. To dig pits for the saplings they'd grown in their nurseries, the Greenworkers had to begin by breaking the ground with crowbars. The transplants were nestled in compost and sprinkled with water that was hand-pumped and hauled bodily. This hard work has produced good results. Reafforestation in Auroville has already improved the microclimate. There are no more dust storms, and the residents say that the shade provided by the trees keeps the ground temperature lower. The effect is most noticeable in the evening, when the ground radiates back less stored heat after sundown.

At the center of the guest complex where I stayed during my few days in Auroville there was a huge banyan tree (*Ficus bengalensis*). What a reproductive strategy! A bird excretes the banyan seed on top of a palmyra and eventually the banyan envelops the host tree with its roots, killing it. The banyan (wild in India) stretches its branches wide, and roots descend from them. These become daughter trees. Single banyans can create huge groves that in time obscure their progenitor, and may cover acres.

In the mornings it was cool—a sweet, refreshing moment after the truly astounding heat of the day. Each morning I awoke to a mounting chorale of bird song, the sound of healing. (The first Auroville bird survey, done in 1972, documented forty species. The next, in 1986, recorded 102.) Loud, brilliant, and varied, this music of regeneration was quite in contrast with the entirely human din of the cities—Benares, Delhi, Madras, and Pondicherry—where I had stopped along the way. I was out in the country—resurrected country—and found it amenable.

Even before tree planting, Aurovilians' first objective was to arrest the hemorrhaging of soil and water into the Bay of Bengal. "In those early years," said Joss Brooks, an Australian who has lived in Auroville since 1969, "the sea was red for ten to twenty kilometers down the coast just with the blood of Auroville." Aurovilians want to reap every bit of precipitation, gathering it into the land rather than allowing it, and whatever soil it can find to carry, to run off into the sea. They began by a practice called *bunding*—the creation of systems of low berms, simple physical obstacles to check and redirect the flow of the ferocious monsoon rainfall characteristic of that swath of India. Watershed restoration in Auroville was learnt by trial and error. When they began bunding, they sometimes followed property lines rather than the contours of the landscape. That approach didn't stop the runoff of water from the land, nor did the construction of large-scale

check dams arrest flash flooding. In November 1978, said Michael
Mason, one of the Aurovilians involved in Greenwork, a sudden, tor-
rential thirty-six centimeters of rain washed away much of their bunding and a costly earth dam. "We learned that you have to start at the
top of the watershed, comprehensively. Year after year, kilometers of
bunding. When we had bunded a large amount of land," he continued, "we could actually go down into the canyons and build
checkdams."

By the mid-1970s, the fruits of some of these labors were becoming evident. "We were beginning to catch large amounts of water with
our bunding and water harvesting techniques," said Greenworker Joss
Brooks. This was a revival of the ancient tank system of catchment
basins, ponds, and small lakes throughout the region that had been
built over centuries to store water for irrigating crops during the dry
season. The system, evidently, worked well, but had been allowed to
deteriorate in modern times.

Because in many places the lands of Auroville had been eroded
down to hardpan, the Aurovilians felt that it was desperately necessary to get anything growing, so they opted to use whatever trees
could survive. Two Australian tree species, *Acacia auriculiformis* (the
ear-pod wattle or northern black wattle), and *Acacia holosericea* (wah-
roon in its native Australia) were heavily employed. According to the
nomenclature developed by Mirra Alfassa, the Mother, *Acacia auriculiformis* is known as "Work." *A. holosericea*, which became widely
used some years after the introduction of *A. auriculiformis*, was
dubbed "Silver Work" by Aurovilians for the color of its leaves. Not
being native to India, they have, like any introduced exotic plant species, the potential to run amok. On the other hand they do make a
good "pioneer" species (the worry is that they'll behave like the pioneers in the American West). In some of the most eroded lands I saw
from the back of Ed Giordano's peppy motor scooter, acacias were
virtually the only things growing. They did afford the hope that some
other species might also come to grow, from seeds brought in the guts
of perching birds, or from seeds borne in on the wind and lucky to

find a sheltered spot beneath the leaf litter. Whatever comes, though, is going to have to be pretty tough to succeed the acacia, which is enormously prolific.

Inadvertently, Auroville popularized these acacias in India. They have been adopted all over the country by forest departments and planted in situations where concern over the indigeneity of the vegetation is likely to be slight. The Greenworkers are aware of the need to contain the spread of exotics, and debates over this practice continue. David Nagel, a Greenworker from the Aurodam settlement who does village development work, says, "I say it can be controlled because I know it's worth money." *Acacia auriculiformis*, the Work tree, produces a respectable lumber, suitable for furniture making. Nagel says, "It's called in the local market 'Third-class Teak.'"

Like its architecture and appropriate technologies, the Greenwork at Auroville has developed diverse tendencies. They say that there are as many different kinds of afforestation as there are people, and that rather than developing a generalized philosophy, they are trying to "find ways of working that work." In Giordano's mind, seeds are the alpha and the omega. "I learned first the nursery stage," he said. "Going back and forth to gather seeds I got to know the forest stage . . . it's like watching your kid grow. Back then I would more easily be able to identify the seed than the tree." In an initial planting for afforestation, these days the Greenworkers plant a mix of trees—dalbergias, teak, rosewood, three to four kinds of mahogany, *Prosopis juliflora* (mesquite), *Azadirachta indica* (neem), mango, and custard apple (the latter, they say, attracts animals, thus "bringing in that work force that will do the job for us").

Auroville has generated its own sylva, booklets titled "Trees of Auroville" and "Common Trees of Auroville," listings which give the botanic names, the common English and Tamil names, as well as the Mother's names for the various trees. The epithets the Mother bestowed on some 800 flowering plants and trees—her divination of their essential properties—make a hymn of the forest: names like Supramental Knowledge (sweet acacia), Integral Wisdom (East Indian walnut), Stability in the Vital (purple bauhinia), Joy of Vegetal Nature in Answer to the New Light (bottlebrush), Material Enterprises (silk cotton), Multitude (the coco palm), Mental Fantasy (poinciana), Intimacy with the Divine (crape myrtle), Aspiration (night jessamine), and Religious Thought (ivory wood).

Out of their experience, the Greenworkers have also developed "The Nursery," a very serviceable handbook with simple, use-tested instructions for propagating trees for reafforestation in the tropics. To deal with the threat the fearsome sun poses to tender seedlings, there are suggestions for providing awnings. There's advice on spacing rows and creating beds underlain by wood ash to ward off termite onslaughts from below—techniques as affordable as they can possibly be, yet still perhaps beyond the means of many.[4]

Here and there in Auroville are a number of small (one- or two-acre) plots in addition to a few larger plantations established as sanctuaries—genetic banks. The plant community that Aurovilians are, in their sanctuaries, attempting to restore is known as scrub jungle. An even more technical, yet more evocative name for the vegetation community of this area, the Coromandel Coast, is "tropical, semi-evergreen thicket." Plants in this climate, where the dry season extends from January to June, are capable of enduring drought either by the strategy of sending down a deep root, or evading it by racing through their life cycle in the rainy season, which lasts from October to December. Scrub jungle is rich—there are known to be 266 plant species in the semideciduous scrub jungle: 104 trees, 59 shrubs, 52 lianas, 36 herbs, 14 grasses, one bamboo. It would consist of a dominant stratum of trees reaching seven to twelve meters in height, including *Albizia amara* (mimosa amara), *Butea monosperma* (bastard teak), *Dalbergia paniculata*, *Pterospermum suberifolium*, *Azadirachta indica* (neem), and *Syzygium cumini* (Indian allspice). Underneath these would be thorny shrubs, shrubby trees, climbing woody vine-like lianas, some herbaceous annuals, and a few grasses.[5] The Greenworkers developed their own rough understanding of the native vegetation complex ("We went through books, poetry, and temple inscriptions to find out what was here," said Giordano). The English did such a job of leveling the woodlands of India that to discover what species the local scrub jungle comprised, scholars have made recourse to temple groves to study some of the last remnants of undisturbed (but not, alas, unendangered) vegetation, where trees 200 to 300 years old, and their full complement of associates, might be found. Now, in Auroville's sanctuaries, "Nature is choosing the right stuff." In some of the forests they have planted in Auroville, natural reproduction has become so successful that seedling nurseries are no longer needed. Whereupon, says Brooks, "forestation becomes more poetic."

My Auroville tour included a visit to the settlement where Joss Brooks lives. Brooks was sick, suffering with a fever, and short on time, but gamely played the gracious host. The compound, which included his cottage domicile and a soaring, thatch-roofed open kitchen, was surrounded by scrub jungle. Regenerated scrub jungle. You might infer from that term jungle a dense green opacity, but the texture of the vegetation was not impenetrable. Rather, there was a laciness of soaring trunks and looping vines and droughty, drooping blossoms, among them the heavily fragrant plumeria. This place in 1973 had been an empty field. Over the intervening decades of reinhabitation, Brooks came to know the members of this plant community in the affective way you come to know your human neighbors, by their strengths, weaknesses, and peculiarities. His style of Greenwork is a reverent, curious experimentation. For instance, as we passed through his garden he mentioned that he was cultivating a wild cashew found in the mountains—and walked on, commenting about plant after plant and their myriad ways. He remarked a "thorny banyan" putting out roots and described its generation of thousands of enticing date-like fruits. He said it crops up everywhere that birds pause to eat and excrete. A rotting palmyra, he said, might house a bandicoot (a giant rat). Thus much of Brooks's observation centers on relationships, which is to say, ecology.

In describing his approach to restoration, Brooks said one could just "look at all the different family relations that can be in an ecosystem, start from all the different levels and continue enhancing." This continuing enhancement was abundantly clear in the forest surrounding Brooks's compound. Even absent much botanical knowledge, I could see that there was a vastly greater diversity of plants in these few acres than were present in the stark, gullied lands elsewhere in Auroville, or in areas just in the earliest stages of regeneration.

Aurovilians respect the will of Nature to regenerate, given the slightest encouragement. Yoga means yoking, and in Auroville, the Greenworkers seek to yoke themselves to the life force, Nature, Shakti, call it what you will—the goal is to ally with it and do the work. "Shakti is Nature's executive force," said Ed Giordano. "We have found that it's the force behind it in all that we're dealing with." Thus Shakti is the lead botanist in Auroville: "She's planting stuff that we wouldn't normally plant; so part of our job is to figure out why." The dynamics are intense: abandon a nursery and "a few years later you find forest giants," said Brooks. Revivifying the Shakti of a place, one gathers, "does depend on some people loving it and relating to it and helping different aspects," as Brooks put it. Meanwhile, the overall pressure on the land to produce salable commodities is so heavy that one of the functions of the forest sanctuaries within Auroville is to permit what Brooks called the "incredible luxury of watching a tree rot."

Auroville's applied ecology grew from the ground up. No one arrived at the work of stewardship with the appropriate technical training, so Greenwork is largely autodidact and experimental, but with a fitting degree of rigor in the researches. The few Greenworkers, a tiny fraction of Auroville's membership, are rigorous amateur natural historians, researching and developing a knowledge base about the land they are making their home. Like the Mattole Valley restorationists, they were able to supply rainfall records, as well as graphs of solar radiation and day length, wind direction and velocity, temperature charts for the area, maps of the plateau's geology, all the mass of detail necessary to understanding the Nature of the place where they're investing their lives and souls. They follow the principle of starting from observation rather than theory. The yield would be an indigenous method of restoration and reinhabitation. Cosmopolitan indigenous.

While touring the sanctuary we crossed paths with Mayakannan, a local man knowledgeable about the plants who works in the nursery here. This interest in plant lore makes Mayakannan a rarity. In another setting, remarked Brooks, "he would have been a beatnik," the kind of guy who would just take off for some months to walk in the hills, learning the natural history, or to sit around the village listening to the old people who still know the plants. "There are few like Mayakannan in the villages these days," Brooks wrote to me the following year. "The knowledge bank of information about medicinal plants is emptying as the old people pass on. Two sites in Auroville have been selected as part of a network covering the three southern Indian states to conserve and regenerate endangered species of medicinal plants. Part of the work is studying and documenting the ethnobotany of the region from the remaining herbal doctors and growing in the Auroville sanctuaries the plants needed for their work, at the same time as generating awareness through the schools and community primary health care groups about the availability and curative properties of the Coromandel Coast flora.

"It marks a significant development in the Greenwork of Auroville, and hopefully the existing pioneer forests will be gradually retrofitted with the truer indigenous species of the area."[6]

In a workshop slide show on the Greenwork, Joss Brooks spoke of different kinds of experimentation with revegetation in the earlier days. Preventing damage to plants from herds of goats and cattle, or from other assaults, was essential: "Sometimes it was protecting just to watch what would happen," presumably to see what healing the elimination of grazing alone would result in. "It's naïve to think of any kind of afforestation work without good protection," said Brooks. It "must be there on an infinite basis." Although "it means giving up some of your land," fences buy time for the establishment of forest sanctuaries and agroforestry. The sisal and thorn fences Aurovilians plant become the same kind of rural landscape niche that hedgerows do, serving as wildlife corridors, long, narrow strips of habitat for birds especially.

The villagers living around Auroville did not initially welcome the reclamation work, either the protection of reafforested sites or the presence of earthen berms to capture runoff. In one instance the villagers came and broke the erosion control work upstream in the watershed because they wanted the runoff for their *kollams*, or tanks. Any benefit to them from the recharge of the aquifer would be virtually imperceptible—a benefit to the landscape and their future, but not an effect that would accrue to the individual landholder in immediate economic terms. Brooks later commented: "Aurovilians learnt that essential watershed management principle: 'Everyone lives downstream.' Dams were built that allowed the first water to flow to village tanks downstream. When they were full only then were gates closed so that larger bodies of water could be stored upstream."

Changing land use patterns in the region have perhaps relieved a little of the grazing pressure exerted by cows and goats, but have brought along the different set of problems associated with cash crops and monocultures. And these are not so easy to deal with at the local level. For restoration in any degree of fidelity to take hold in the developing (and, our probably de-developing) world, it must somehow acknowledge human economic need. In India, in all but the most exceptional cases, this implies agroforestry, requiring that the regenerating landscape provide a harvest of food and fuel. Agroforestry is a longer-term proposition; and with the galloping trade in commodities and luxury crops, rural economies everywhere have been put on fast forward.

The biggest challenge facing restoration activity in Auroville, or anywhere, is to get it to make sense in a civilization whose relationship to land today is almost entirely exploitive, with parks and preserves (of which India has established quite a few) being the exception in a pattern of land use for profit rather than basic subsistence. The fact that authentic land healing is not considered "economic"—in the sense of being remunerative—is an exact measure of the worthless malignity of the "free trade" economic paradigm, which has yet to take into account that degraded land is both economically and ecologically unproductive. Land health, in its fullness, is an incalculable good. For now, land healing is mostly volunteer work, mostly donated. It only makes sense in the longer view. To make public policy take the longer view, as it only occasionally does, requires a tremendous amount of public education and a profound change of heart.

Aurovilians' struggles to capture rainfall and maintain the aquifer go on in an area where the state subsidizes electricity for agricultural uses. The well pumps run twenty-four hours a day and groundwater is being lavished on thirsty crops such as sugarcane, not because sugarcane is in any sense ecologically appropriate to the region, but because the peccable logic of the market currently recommends it. As a result of this profligacy, said Mike Mason, there has been a six-to-eighteen-meter drop in the local groundwater level in the last sixteen years.

All over India the Green Revolution has worked its lethal distortions of agricultural practice. As part of cultivating cashews which are a major export crop, says Brooks, during nearly half the year "the most incredible pesticides are sprayed by people half-naked, with no protection whatsoever. Nausea, sickness, and even death are not uncommon." As more land is being used for export crops, less food grain, principally the traditional local millets, is being grown for local consumption. The Green Revolution, abetted by the government policy subsidizing electricity for groundwater pumping, has spurred the expansion of irrigated agriculture. There is a real possibility that groundwater overdrafts could, by inviting saltwater intrusion, literally kill the aquifer that the Auroville experiment and its surroundings depend on.

Auroville is by no means food self-sufficient, although that is among its goals. India once produced tens of thousands of locally adapted varieties of rice, as many as there were villages, probably. This tremendous diversity has been overshadowed by the one-size-doesn't-quite-fit-any International Rice Research Institute rices that yield a greater volume of rice grains per acre, but which require fertilizer and pesticides, are less complete nutritionally, and produce less straw, hence less sustenance for the draft and dairy animals on the farm. Auroville's organic farmer Bernhard Declercq has collected and is keeping in production many of the local varieties of rice. These and other experiments in organic agriculture are being pursued on Annapoorna, a 135-acre tract of "black cotton" soil, which is as unforgiving of abuse in its way as the bricklike laterized soil elsewhere on the plateau. Annapoorna is salinated dryland, soil that has accumulated a burden of mineral salts as a result of irrigation. The hope some day is that Annapoorna will provide food for Auroville. The reality is that they're still trying to find ways to produce a yield with the minimum energy input it takes to run a rototiller and with organic soil amendments only.

The confusions and losses wished on all cultures by industrial civilization are tremendous. Civilization's history, it now seems, has been a war upon subsistence and the creation of economically privileged elites. The colonization of India exacerbated this trend mightily, and the new imperium of transnational capital is unlikely to go a different way. The spiritual experiment in Auroville is going on in real time and it is unsurprising that it has yet to produce a real social equality in its broader surround. There's an enormous gulf between Auroville's eco-futuristic lifeway and the slow, hungry, rudimentary quality of village life, life reduced to bare survival by some centuries of basic oppression.

Labor is extremely cheap in India—a day's wages for a landless laborer amount to slightly less than a Yankee dollar. A skilled laborer like a carpenter or a mason may earn about $1.80 per day. Nowadays almost all the manual labor in Auroville is hired—the work in planting and building the bunds, among other activities, has employed more than 2,000 villagers. "We're supplying employment," says Giordano. "Most villagers don't own land anymore anyway." Women nursery workers in Auroville receive a little more than fifty cents a day. So help is not hard to find and in the few households I glimpsed servants were the rule. Ed Giordano defended the practice, saying, "Most of us spent years out in the fields," and that "today there is far, far too much to do and too few of us to do it."

The quality of life for Aurovilians (which is simply adequate by U.S. standards, including, as it does, a sufficiency of food, household water enough to run the occasional washing machine, electrical illumination, a reasonable measure of cleanliness, cooking fuel in ready supply) seems positively opulent in comparison with the low-smoky huts of the mostly impoverished Tamil villagers in and around Auroville, with their sense of stasis or retrogression.

Yet the Tamils are the first people of the region, and, in a sense, of the subcontinent of India, which had its Columbus day about 3,500 years ago when the pale Aryan invaders, with their worship of fire and air, and holocausts of the forests harboring tribal peoples, came down out of the north to overpower the darker Dravidians, who were the original inhabitants of the subcontinent of the south. The Columbian Quincentennial, with its antiheroism, provided North Americans with some overdue correction to the prevailing romances about exploration, colonization, and the presence of indigenous people. In 1992 the Western Hemisphere was informed, by Fourth World peoples, that it was not so much a "New" world being discovered in the Indies as a myriad of fairly stable and enduring tribal peoples and their ecosystems being swept away; that the so-called New World was to industrial civilization a never-to-be-repeated plunder of land and wealth. Consider the contrast: on arrival in Auroville, said Mike Mason, "we wanted to heal the land, to bring back the life of the land." This necessary intention, the antithesis of colonial exploitation, represents a turning point in human history. The next Terra Nuova will be Planet Earth, restored.

Auroville is not all that unlike the rest of the world, especially the developing world, in the messy, unequal, and unresolved relationships of its peoples. There are westerners descending with their advantage of hindsight about centrist, industrial development, who as devotees of Sri Aurobindo and the Mother also have a syncretic kind of foresight into the human potential. And there are Indians with their second sight, the people who have dwelled here timelessly, and whose lives of late—especially in the last 300 years—have been washed over by the various tides of "progress." The British mined the region's forests, and since independence India's own cash economy, under the tutelage of the World Bank and the Nehru dynasty, is mining the people. The brain drain runs from village to city to overdeveloped country offshore.

The westerners in Auroville seem to be meeting Third World poverty halfway. They have reduced their consumption tremendously from what it would have been in the United States or Europe. In addition to restoring groundwater, soil, and scrub jungle, Aurovilians are mindful of the need to restore at least a precolonial quality of life—that is, ample subsistence—to the village, but the situation is dauntingly complicated. And it is a kind of microcosm of the world with the West in it. The ways of arriving at consensus toward reinhabitation in populations that are heterodox in every respect—culturally, economically, linguistically, educationally, in terms of the status gender determines, and in terms of land tenure or lack thereof—can only have some hope of succeeding in such microcosms, at a certain scale, and in specific places.

Auroville possesses a number of the elements that, I think, must be required for sustainability in our time, and into the future. Above all, there is a great vision. Auroville is not merely a reaction to present problems, but an experiment in creating something more beautiful, coherent, earth-healing, and soul-satisfying than people have ever known before. "Ideally, as originally conceived," wrote Ed Giordano, "Auroville should eventually be a city of 50,000 so as to provide a certain critical mass for cultural, cosmopolitan, urban, and other phenomena to be able to happen. Greenwork has till now been our most important and conspicuous work and will remain an ongoing challenge to refine and expand. But the present and future challenge is to integrate it fully into the urban dream for Auroville. This is what Auroville is—bridging the impossible (very difficult) into reality. Transformation is our present work—transcending this situation and manifesting the next phase of evolution is the real work.'"

A decade ago, mainstream media like *Newsweek* and the *Los Angeles Times* were predicting the community's demise because, the pundits figured, Auroville's ideals of human unity were not only impossible of achievement but that idealism itself constituted a fatal flaw. Certainly Auroville has a distance to go in attaining unity, but it is earnestly moving in that direction. Despite some turmoil following the Mother's death, and the worsening planetary situation, Aurovilians persist in their experiment, and with an attractive sense of mission.

"We are learning," writes Aurovilian Bill Sullivan, "that the normal economic and social incentives are not enough to motivate toward sustainable development of the planet. . . . Auroville has been able to weather the storms because the motivations of the participants in this 'laboratory of evolution' are precisely based on such idealism."[8]

Chapter 10

Devotional Exercise

You can run, but you can't hide. Whether you're an oven-bird, an orchid, or a middle-aged divorcée, sometime, somewhere, you've got to make your stand, declare a tie to the ground you're standing on and to the larger community of the land. One day in the spring of 1993 my neighbor told me that the landfill that lies in the section due south of here had applied for yet another expansion, which would mean not just a lateral but a vertical increase. Thus I might someday look across the back forty and see a tumulus of trash ten stories high. No telling how much denser the flocks of herring gulls, battening themselves on an even larger offal buffet, would become. My neighbor muttered darkly about selling the house that she and her husband had hand-built, to get away from the stench.

I too briefly weighed the possibility of fleeing, of getting out while the getting was good. But eventually the thought of what might happen to the land if I weren't around to protect it overtook me. Psychologically speaking, I have evolved into a niche specialist. It turns out that my habitat requirement is very narrow. I cannot imagine a life that doesn't include a symbiotic relationship with my old beech tree.

Or, more precisely, I can't live with the idea that somebody else might decide that old tree was "overmature" and inconveniently occupying a nice site for a pole barn.

That old beech is sacred to me. Today, late in the winter of 1993–94, I skiied out to visit. The tree is magnificent in every season. Impulsively, out loud, I vowed that I would do everything in my power to see that the tree could live out its days. Then I returned to my studio, picked up a conservation biology magazine, and read about the possibility that insect pests like those that all but eliminated the chestnut and the American elm also have begun to infest beeches, dogwood, hemlock, and many other North American tree species; I read a prediction that a few degrees' rise in temperature resulting from global warming could eliminate the beech, paper birch, and sugar maple, as well as scores of other forest trees from the United States within sixty years. When I picture the desolated landscape that might result from these depradations, the remark about the horror of nuclear war—that the living will envy the dead—creeps into my mind. Being reapprised of the magnitude of the continental and global threats to the living beings of the uncounted species that constitute the big abstraction of biodiversity where I live made me realize that in vowing to protect the beech tree, I had taken on a purpose that dwarfs me. The only hope of fulfilling the vow resides not in my meager being, but in devotion to something sacred in Nature, something greater and more mysterious than I can frame.

Such devotion, I have seen, can lead people to take on seemingly impossible causes. It moves them beyond despair and into action. Traveling in India I had met a Hindu priest in Benares who ten years before had espoused the cause of cleaning up the Ganges River. To do that will be about as easy as preserving and restoring the Eastern Hardwood Forest. Yet once the vow is made, it is no longer possible *not* to act. What's more, help may come in fateful ways. In India I was a visitor in an old, old civilization; a civilization overlain on even older aboriginal cultures at whose base was animism—the belief that groves, pools, streams, and serpents also have spiritual power and can appreciate devotion. Persistence of this understanding was evident in the spontaneous veneration of certain termite mounds in Tamil Nadu (they are seen as manifestations of the generative aspect of the great god Shiva), and in the worship of certain trees and stones as divine— no consecration necessary. Thus devotion to the Ganges is a persistence from prehistory, not a relic, but the unbroken continuation of the worship of wild Nature, of rivers and mountains, beasts and birds, seas and storms.

The city of Benares is old, one of the oldest living cities on Earth. It had been such a numinous place for so many centuries that the Buddha saw fit to launch his preaching career from its vicinity. *Banaras: City of Light*, Diana L. Eck's marvelous book about the city, details its sacred geography and archaeology, but there is no definitive answer as to why this particular site accreted such *mana*. The natural setting surely had something to do with it, primordial qualities of landscape that vanished into beyond-dust centuries ago. The place that is Benares once was a forest bejeweled with pools. It sits on the higher bank of the Ganges at a point where the river recurves north. It was a sublime place for human habitation and, evidently, a propitious place for divine visitation. Here and there, isolated scions of that forest, great *peepal* trees, persist amid the density of human settlement. They, too, symbolize the life force and also are venerated. Benares is sacred to Hindus, a destination for millions of pilgrims. To die in Benares secures freedom from the endless cycle of rebirths. It is a consummation sought by devout Indians who come at the close of their lives from throughout the country, hoping to expire in this city sacred to Shiva, the god of destruction and creation, and to be cremated on the river's edge. Their ashes then are cast into the holy—and now appallingly polluted—Ganges.

I first learned of Swatcha Ganga, the Ganges clean-up effort, through a Traverse City engineer. Jim Porter and his daughter Catherine had both contributed their skills to the international citizens' effort to support Swatcha Ganga, which is headquartered, fittingly, in Benares.[1] Being somewhat of a devotee of seemingly impossible causes, I was interested in meeting the Ganges activists. I also must confess that Benares' antique charm and religious significance drew me, too. Perhaps impossible causes become a little less so by virtue of apparently incidental attractions that may in truth be the subtle animating essence of the work.

The head of the Sankat Mochan Foundation, which is the auspice for Swatcha Ganga as well as a number of cultural activities in Benares, is a high priest, or mahant. Dr. V. B. Mishra, or Mahant-ji, as he is known to his fans, oversees the Hanuman temple in Benares. (Hanuman is the monkey god who epitomizes devoted service to the divine in India's epic *Ramayana*.) The Mahant is also a professor of hydrology and civil engineering at Benares Hindu University.

The love of the mother Ganges is a wellspring of the Hindu faith. The depth and breadth of feeling that the Mahant, Dr. Mishra, would exhibit as he talked about the Ganges, made my Earth First! compañeros seem like dilettantes by comparison. Assuredly they are not. The idea that the Ganges is divine is not, alas, protection against the terrible contamination of the sacred river itself, and by practices such as to give a westerner the willies. Further demonstration, if any were needed, of the limits of the ability of a spiritual system to govern temporal affairs. Human logic has an amazing torque. Thus there are Brahmins who hold that the sacredness of the Ganges means that the river is, by its very nature, unpollutable, beyond harm by mere mortals.

The Mahant is not one of these. My preconceived notions of Dr. Mishra were all wrong. Because he is an engineer, and because he and his associates in Swatcha Ganga spend a lot of time running tests on the water quality of the Ganges, I had assumed that he would prove to be a technology dupe, fascinated by management and involved in what could be accomplished scientifically, that he'd have the engineer's can-do hubris. In this as in most things, India overturned my surmises. With disarming humility he said, "I am by virtue of my training a member of the community of polluters." His leadership of Swatcha Ganga would argue otherwise. He is a very busy man, being occupied with both his temple duties and his professorship at Benares Hindu University.

The aims of Ganges restoration are essentially human-centered, hardly ecological in the sense of restoring a larger community of organisms, or the quality of an original riparian system. When, during our interview, I asked the Mahant whether there was any concern about the fauna of the Ganges he found that a sort of risible proposition, a luxurious idea. His passion is to restore the Ganges waters to a degree of wholesomeness such that devout Hindus wishing to take their daily holy dip in the river need not flinch at the visible pollutants, nor dread infection by microorganisms. As he spoke, he conveyed a palpable sense of the devotional quality of his work to stop the pollution of this holy river. In response to a question about the river as a source of waterborne disease, the Mahant said that the least that the welfare state of India could do would be to provide the millions of devotees to Ganga with a river fresh and pure. "Let them drink that water that they consider nectar and divine," he declared. I think he regarded a reductionist approach to the river—dealing with it as a health hazard—as anathema. V. B. Mishra seemed to feel that the river's holiness should be reason enough to do everything necessary to clean it. He abhors the thought that the caution against waterborne illness might alienate people from their mother and was magnificently outraged at being as yet powerless to prevent sewage from being "smeared around her face"! One got the feeling that for the Mahant, this was no mere metaphor.

After his *cri de coeur*, Mahant-ji continued: "If there is no love and commitment in this world, then this world will disintegrate like sand particles. The cohesive force in this world is love and commitment—the only cohesive force between all of us is this. . . . That is what we should be living for on this Earth, otherwise there is no need to be living on this Earth." This compelling assertion of what we are here for moved me to tears. When Mahant-ji noticed my weeping, he said to me that he, too, would cry about this, over and over again.

When I spontaneously uttered my promise to the beech tree, it was unqualified, although it may have escaped my lips easily because at first I meant simply that I wouldn't let anyone cut her down. To desecrate that tree would be an affront to the memory of the maple-beech forest that once stood where she does and a breach of faith with the maple-beech forest that may yet, against all odds, grow up around her again. Although the probabilities are looking grim, the possibilities remain to be pursued, regardless. What else is there to do? The one set of odds that I have the greatest hope of influencing is determined by my own human limitations, the way I tend to think and act when there's no mahant around. Executing one's good intentions is ever the trickiest part.

At times my relationship with my land has been quite as judgmental as my relationship with some significant human others in my life. I have "known" it to be damaged, impoverished, manipulated, and not the best. I have coveted my neighbor's woods. I have felt mortally threatened by all these wretched, marginal pines, many of them sick and dying, not irrationally fearing that in a drought year it wouldn't take much to kick off a good old-fashioned Michigan forest firestorm. From time to time, I have resented the sense of obligation to do something about the wounds and Earth-afflictions, being in my essential character a sedentary type now facing a vast thirty-five acres of yard work. I envy those whose love of their land is unalloyed. Maybe I am cursed by having as yet mainly restoration eyes, but no restoration skill, back, or upper body.

Despite watching weeds encroach, sand blow, pines pine, and fire hazards mount, I love visiting the patchwork of places out back of the house. Spectral dead pines grizzled with lichen. Stemmy, deer-browsed young maples, nuzzling right up to the Scotch pines for shelter. Small, secret clearings. Blackberry thickets and popple copses. Every ramble through the property reminds me that there is a universe-a-plenty here, with not a square foot unworthy of respect. Affection returns and I'm forced to conclude that the whole community has got to be held sacred, even the nattering of my thoughts, or nothing on Earth can be.

One wet day in August 1991, I bestirred myself to follow a suggestion from Malcolm Margolin's wonderful *Earth Manual*, which is a guide to restoration projects for individuals.[2] He urges the would-be erosion battler "to try to forget everything your mother ever taught you about 'catching your death of cold' " and to go out in the rain and see what the water is doing on the land. Despite the downpour, it was a mild day, so I figured I could risk a soaking. I started threading my way back through the thicket, and soon was sopping wet, squelching around dreamily *sans* spectacles, which had been rendered useless by fog and droplets. I looked up and saw the rain falling at me out of a blotter-flat sky; I watched it course in rivulets down the trunks of the chokecherry trees.

Emerging from the pines to the south slope of a little knoll whose north face I regard from my writing studio, I saw the makings of gullies. White pine seedlings planted there a couple of years ago weren't quite able to hold the ground. That summer's periods of drought alternating with slashing rains ensured that until I figured out something to do about that raw hillside, the sound of rain on the roof would just be tributary to a Mississippi of eco-guilt. A few mornings after the rain-soaked reconnaissance, some more of Margolin's advice registered and I concocted a way to build some brush dams with available materials. On either side of the incipient gullies I pounded in some sharp sticks. Then I lopped branches from the omnipresent pines, thanking the trees for their sacrifice to the cause. I wove the branches crossways against the stakes so as to impede the water's flow and catch the soil on its way down. The hope was simply to prevent a blowout, which is how great ugly gashes in the land are referred to in my part of the country. On my land, ecological restoration is yet a ways away.

A year or so later, during a study break, I walked down the little hill atop which perches my writing studio, across the lush pastures of the septic system's drain field, and up the scrofulous slope beyond. The condition of this small scrap of my holdings is very disheartening. It has been losing vegetative cover ever since I arrived. It started out with a crust of lichen, but the passage of feet has broken its surface, opened it up to the ravages of drought and rain. My former husband and I tried to get some walnut trees started on the north-facing slope, but they died after a year. On the other side, my funky brush dams of the previous year were meeting with mixed success in arresting the gullying process. Behind a couple of them some terraces of sand were building, upon which I think I may have spied some knapweed taking hold. Not a great result, but commensurate with the amount of labor and study invested.

Mostly I theorize and rhapsodize; in extremity, I act. I had a bag of rye seed lurking around the crawl space under the house, so I dredged it out. It dated from a few years earlier—some of it had been the cover crop for the inaugural garden. Doubtless it had long since lost its capacity for germination, and being non-native, feed-store bought, was certainly no candidate for restoration seeding. I don't yet know what grasses might be native to openings in this area. And if I did, could I find some seed? And if I found some seed, could I afford it? And if I planted it, could I tend it? And if I tended it, would it flourish? Speculative questions, all, that may await empirical response for a long, long time. What the rye had going for it was being here. I was willing to invest in some conventional lawn fertilizer to encourage it. For the next few weeks two five-gallon buckets sat in my studio, one half-full of seed, one of fertilizer, and from time to time I would haul them over to the slope and scatter handfuls from each, rubbing them into the exposed sand with my bare feet. That was my restoration work—very desultory—for early July. Probably July wasn't the right time of year for it, and the next day a crow came along to snack on the seeds. I watched the crow with no grievance, only the hope that the bird would be clever enough not to ingest some of the chemical granules along with the stale rye seeds.

Messing around with the seeds, and realizing the scale of effort unfolding before me, brought Psyche's labors to mind. You know the myth. Invisible Cupid woos Psyche, spirits her away to an Olympian love nest, forbids her to look upon his face, comes to her by night. Relationships between mortals and gods are always trouble-fraught. Psyche's sisters arouse her curiosity, undermine her trust and happiness. As Cupid sleeps, she peeks. A drop of hot oil from her lamp awakens him. He casts her out. The stern mother-in-law, Venus, sets the terms of the reconciliation: impossible tasks, among them sorting out the mixed seeds in a large granary. Psyche, overwhelmed by the enormity of the job, swoons in despair. While she sleeps, a myriad of ants comes and attends to the sorting. Where, I ask, are my ants? I'm ready to swoon.

As it happens, ants are essential to the life of the hardwood forest, and effect their contribution by carrying seeds. During our interview at the University of Wisconsin at Madison's Arboretum, the ecologist Virginia Kline mentioned that the wildflowers and low-lying plants of their restored beech-maple forest were failing to reproduce. The trees were returning, but, owing possibly to the dearth of ants that had resulted from earlier logging and clearing, the herbs weren't regenerating. These plants and ants had co-evolved. The ants were enticed to distribute the flower seeds by edible coverings on the seeds that also functioned as handles for toting the seeds down to the ant colony's storage chambers. Uncanny that evolution found a way to disperse these seeds, whose germs remain intact after the ants have their feast. The ants "rescue" the seed from other forest-floor consumers, plant it at a right depth in the earth, and in beds well supplied with ant guano.

The moral of both the Hellenic myth and the ecologist's anecdote is that the seemingly insurmountable task of restoration will be effected, ultimately, by tiny beings: pollinators, mycorrhizae, root hairs, spirit helpers. All volunteers are welcome. Out back I do notice maples and beeches volunteering hopefully amid the pines. It appears that the undying spirit of the hardwood forest lurks like a revenant in the pine thicket, and that with a little help from its friends, it could reclaim its former ground.

Despite some sedentary tendencies, I have learned that the eventual outcome of a protracted spell of indoor inactivity is an almost terminal funk. Eventually, I bestir myself. On a mild day in early December 1992 the sun came out. (In cloudy northwestern Lower Michigan, this can be a noteworthy occurrence.) There was snow on the ground but not so much as to turn a walk into a slog. The previous morning, as winter's malaise settled deeper into my psyche, I had realized that the reason my conscious contact with Nature was so feeble was that I hadn't been out of doors in a serious way for weeks. I had the hermit's version of cabin fever. I was tiring of my own personality for company, and of only human constructs, even of my prayer life, for spiritual experience. A surfeit of abstract worship and petition, but no epiphany. I'd fallen into the error of too much efforting, not enough surrendering. Enough of that and everything becomes foul with sweat. So I had promised myself a walk in the open air.

Qualifying the appeal of the walk, however, was the matter of my intention to restore the woods. It seemed as though I should be doing something back there to make an honest woman out of myself. In truth, for me to begin the work of restoring my acres to postglacial mint condition, I mostly have to kill Scotch pine trees. People find the idea of restoration lovely because it often involves planting, especially planting trees. Much restoration work, we now know, entails the removal of exotic species. Rick Moore, the forester, had allowed as much one winter as we tramped through the pines, surveying my property with an eye to bringing back the hardwoods. "Just get back in here and start cutting," said he. "Winter's the perfect time to do it."

Winter of 1991 had been my winter to learn to walk in the snowy woods. Simply getting out there and developing some ease with the cold and the snow was a prerequisite. By December 1992, I had some confidence in my ability to navigate in these conditions. No doubt this degree of trepidation will seem excessive to readers who've grown up in snow country. Permit me to explain. Not only did I come into snowy terrain at about the same point in life—the late thirties—that Dante came to himself in the middle of the dark woods, having lost the straight path, I did so on a broken leg. As the result of my automobile accident, I spent nearly four years with one leg vulnerable to re-injury by a fall or a jolt. In fact one winter I did break a prosthesis that had been implanted in my leg—a flawed piece of fancy steel alloy—by crunching down through a crust of snow. So it has taken me a few years to overcome the fears stemming from those experiences and conditions which are now healing and healed, respectively. Quite apart from that individual circumstance, there's the general truth that to learn new terrain and its habit takes a long time, maybe more than a lifetime.

This day's sun was inviting enough, and the urgency of making good on this project compelling enough, that I suited up, booted up, and sallied forth with my bow saw and loppers, and my short Michigan snow leopards, Simone and Tyrone. "Mills's Hardwoods Liberation Front," I dubbed myself, walking out through the garden and picking up a deer trail.

Refreshing my sense of the animals' presence was certainly part of my spiritual practice that afternoon, restoring to my senses the awareness of the other kinds of lives and other destinies that are tangent with mine by virtue of our sharing this home. However shared our territory, the deer, the coyote, the porcupine, the cottontail, the jays and chickadees all fell silent as our little party, consisting of three of their worst predators, passed through the thicket. I was just looking for some feasible situations, healthy hardwoods surrounded by pines of a size I could safely fell. What I aimed to do was liberate some hardwood saplings from the shade of the conifers. In the green world, life is about competition for light. Plants have evolved to grab light at the most propitious time, and in sufficient measure. Understory plants may make their bid quickly, in the springtime, while the tall trees are still bare. And the countless seedlings of spring get theirs throughout the summer, as the breezes and winds stir the canopy, allowing sunflecks to dance below. Hardwoods want light, being not infinitely shade tolerant, and the pines provide year-round shade. So we foster hardwoods by lopping away the pine branches that obscure their broad leaves.

Lopping branches is one thing, taking down whole trees quite another. It's just barely within my competence to use a bow saw at all. Felling trees, even the ten- and fifteen-foot striplings that pack my place, has its dangers. Tom Deering, the most experienced woodsman I know, approaches his business of tree surgery and the occasional spot of land-clearing with extreme caution. My neighbor Lowell, a careful man, is recuperating from a broken leg which resulted from a miscalculation about the geometry of a beech hung up in two maples. Lowell had worked in his woods, solo, for years before hurt befell him. His mishap was very much on my mind as I scouted for likely victims. Being alone exacts some compromises, and one of them is that you stay well inside the limits of prudence.

When Tyrone led me off the trail to my second maple hostage-to-be-freed, I judged that I could manage to cut down the Scotch pine that shaded it from the south. I did, but with a mounting panic when I realized that my final cut had caused the tree to fall to the southeast, its top branches meshing with its neighbors'. I had doomed it, but by leaving it hanging I was possibly dooming myself on my next walk through the woods. The cautionary name for such hung-up limbs is widow-maker. I figured that if I made another cut about a yard up the trunk, the top could be guided down onto the ground and out of harm's way. I did, and it worked. Anyone else might have considered my feat to be just a notch above hedge-trimming. However, I didn't fancy becoming the victim of a freak accident and having to discover whether I had the moxie to drag myself through a quarter-mile of snowy, fallen timbers.

Possibly the hankering to thin and weed and control is an atavism. Even hunting and gathering people would burn one set of plants to give precedence to another. Our paleolithic grandmothers, who figured out agriculture, had to exclude one ilk of plants to promote the domestication of others—the roots, grains, fruits, and vegetables that we've all come to depend on. Thus our drive to manage the landscape may have dawned in the garden. Gardening, like restoration, is a learning process, with large lessons in small plots. A major lesson is that the weeds are always at the ready, and in abundance. They respond well to disturbance and cultivation is disturbance par excellence. Weeding may in fact create conditions favorable to weeds.

My dabbling in my garden, and its indifferent results, put me in touch with some ecological truths, grossly stated, caricatured by domestication. Our preferences for plants are narrow and manifestly self-serving—we're hungry, after all. Our objective is to simplify, not diversify. Straight lines make gardening easier. A very few ancestral plants gave rise to the great variety of vegetables we enjoy. By maximizing in these plants the very qualities that also make them attractive to insects and rodents and birds, we pit ourselves in endless competition with adversaries infinitely more numerous and determined (if less destructive, on balance) than we. Protecting crops is either labor-intensive, as in posting guards in the cornfields at night as the moment of ripening causes coons to draw nigh; or capital-intensive, lately calling for fences and poisons. I'm awed by the amount of coaxing and cajoling it takes to grow any food, let alone enough to live on, let alone enough for all of us to live on. And food offers an instant payoff. How then to make commonplace the painstaking horticulture demanded by ecological restoration, which benefits no *one* with a nourishing morsel, but may, by virtue of maintaining biodiversity, help keep the whole ecosystem afloat?

On Easter Day of 1992 I concluded the first installment of a semi-annual rite, the Planting of the Bargain Trees. This is not, strictly speaking, ecological restoration. I'm not sure what it is. Superstition, possibly. Twice a year our local Soil Conservation District offers quantities of seedlings for sale cheap to landowners. The selection of tree species is limited, with heavy emphasis on conifers, red and white pine especially (native here, but not prevalent), and also the spruces. Sometimes they offer red oak (also native in certain places around the county). I have planted them, too. At other times, they've sold red maple, which you might be more likely to find in a swamp than on terrain like mine, sugar maple being more characteristic of these woods. Not to put too fine a point on it, I'd bought fifty red maple seedlings to plant. Easter morning I headed back into the pines with my spade and my cat to do some obscure, perhaps quixotic, landscaping. Chances of those red maples surviving are slim. I don't know where they arose—they were probably raised in a nursery a few hours' drive to the south, fertilized chemically and protected with pesticides —given every modern advantage, except local knowledge. Knowing all of this and going ahead and planting them doesn't make it right, just as confessing that you're a coward doesn't make you brave. Honesty is a good thing, but not the whole thing of goodness.

Of the million Scotch pines out back, at least half impeded my movement through the woods that day, their brittle branches clawing at my braids and swatting me in the chest. Yet I found myself again loving the spaces they create and the privacy they afford me. No one will find me back here, I said to myself as I crashed through, looking for damp spots that might sustain the red maples. I like to think that the red oaks and red maples I've planted among the pines are sending a hardwood-cheering message to the maples and beeches that are recolonizing the ground from their redoubts around the perimeter. The twenty-five nursery-grown seedlings that I planted during my Easter meditation on soil horizons and snow patches are a homeopathic dose in thirty-five acres. Whether or not they survive, I hope the gesture is not lost on the earth around.

Doing the work makes palpable some of the ecological concepts that hitherto lay flat on the page. Under a stand of aspens (locally called popple) there is already a layer of soil forming beneath the leaf litter. The color there is ebony, as compared with the podzolic burnt umber of the sand topped by a mat of needles. The aspen's catkins are emerging, beginning to dangle from the high branches, up out of reach. The coast-to-coast American tree, aspen prepares the way for the climax forest to come, shading itself out of prevalence, colonizing other waste places. No need to tinker with these goings-on; the aspen is a successional restoration expert.

May be that all I'm doing by my labors is continuing a free lunch service for deer: the Whitetail Diner with a Green Plate Special. My pines are a deer haven, and there are enough deer sheltering back there that their trails could be the makings of an erosion problem. Some herbivores—possibly deer, possibly porcupines—keep many of the aspen shoots gnawed down to stubs. If deer, their inordinate presence will work against the reassertion of the hardwoods.

Some days, sitting in my studio, which my hunting friends covet for use as a blind, I see deer in the distance. Perhaps it's the same three or four taking refuge here, coming out to graze in the south end of the valley at dusk. Problematic as the numbers of white-tailed deer may have become, one can only be awed by their grace, by their fleetness, and by the fact that they are sustenance, a longtime gift of good meat, hides, antlers, and sinews to work with, hooves for rattles and adornment. Bearing this in mind, I silently salute the deer—it's not their fault that all this woodland edge was opened up at the same time the timber wolves, bears, and cougars were decimated or extirpated. The deer didn't ask to become a superfluity on the land and a menace to its health. But now their numbers are out of whack, like the numbers of Scotch pines. Unlike the pines, the deer know they're hunted and are magically evasive.

Once, not far from the end of hunting season, a young buck made his conspicuous way up toward my studio, threading in and out of the pines, crossing the path, circumambulating the little building, all unawares of being watched. When he stood among the cherries ten paces to the west, I opened that window and tried to speak to him, chuffing as I had heard deer do. A bad imitation, evidently. The deer was off like a shot across the neighbor's land, bounding away, flaunting his snowy white plume of a tail, a pennant trailing a clean escape, up the rise and into the woods beyond within seconds. Best not to trust a human at any time for any reason.

The day for the 1993 spring ritual tree planting dawned bright, clear, and insistent. The Soil and Water Conservation District sale seedlings had been sitting on the front porch for two days, gasping for dirt. If I did not get them in the ground this day, I would probably be writing off what had been a pretty significant purchase. This year the District Forester had ordered a number of native shrubs, species that you might expect to find in a deciduous woods—witch hazel, serviceberry (a.k.a. shadbush, which harkens to the watersheds where the shad's return is a confirmation of spring), and hazelnut. There were sugar maple, hemlock, and black cherry for native trees. I'd got about ten apiece and planted the shrubs right away, and fairly close to my house, and my studio, especially wanting to keep an eye on the witch hazel which, the books say, *shoots* its seeds when they are ripe. This native tree and shrub planting had a different character from the other plantings I'd done, of species unlikely to be found here spontaneously. It felt more like real ecological restoration. By setting out these members-in-good-standing of the hardwood community to try to establish themselves back in the conifer desert, I hoped to give authentic succession another little nudge. The weekend before, the ground had been prepared for this nudging by some robust young spirit-helpers.

Inspired by the organizing feats of the stewardship network down in Chicago, I had, on the previous weekend, mobilized some volunteer labor, six lively, bright teenage girls from the Northern Michigan Environmental Action Council's youth group (they style themselves the Eco-Spuds, don't ask me why), along with some assorted adults. Tom Deering, the tree surgeon, gave a practicum in the Safe Falling of Scotch Pines. Once the Spuds had accepted the notion that you could improve the woods by killing trees, they took up their bow saws and were loath to put them down. After a few hours spent in two separate crews, threading our way back into the pine thicket and liberating such hardwoods as showed some promise, we rendez-voused at the big old beech tree which was instantly festooned with pretty, happy young women.

This week, then, as I wandered back through, I followed on their handiwork at clearing by planting maple trees in the newly opened patches of sun-warmed ground, wishing the little trees good luck against long odds—sandy, acid soil, droughty summers; deer, cotton-tails, mice, and porcupines all looking to assuage their hunger when winter wears on. Maybe I can persuade some future group of students to help me institute protective measures—wrap the trunks in flexible, inedible tape, hang a redolent bar of toilet soap from a branch to offend the dainty noses of the deer.

As much as anything can, tree planting ties one to the future. As I stood on a rise surrounded by the contorted skeletons of dead and moribund Scotch pines, some picturesque and hoary with lichen, I realized in my gut that this had once been a hardwood forest, that the steady roll of this topography would have been blanketed with a spongy layer of fallen leaves, a tender garden of wildflowers just starting to emerge—hepaticas and trout lilies by now, surely—and throngs of massive trees, still some weeks from leafing out. Perhaps these weekends' work might foster the return of such a forest in the twenty-second century. The idea of leaving my land healthier than when I found it—which, after all, is nothing more than common decency—seems mighty compelling.

I am not unmindful of my privilege in having an extra large and rangy backyard peopled with deer, skunk, grouse, jay, knapweed, the venerable beech, and revelations unexpected—a fox's skull, an indigo bunting. It is more land than I ever would have dreamed of knowing, but I will never be able to disregard its woundedness or cease wondering whether my modest, homespun efforts at restoration won't be obviated by the ecological degradation humans are producing on the planet. I will cry about it again and again, but half the time, at least, those upwellings will be of gratitude for, and pleasure in, the company of these acres, and for the privilege of my modest service to them.

Epilogue

Whereas I write *a poem by dint of mighty cerebration, the yellow-leg* walks *a better one just by lifting his foot.*

—Aldo Leopold, *A Sand County Almanac*

Healing is real, that much I know. It is also a slower process than one might wish, and the rattling impatience of our time outruns it. My body bears the scars of a big injury that befell me when I was thirty-seven. At forty-five, I still notice gains and improvements in the strength and flexibility of that leg that got crushed in the car crash. And every time I notice that, I'm brought back to marvel at this natural phenomenon of health—the capacity of a living system, given favorable circumstances, for self-regeneration.

A few years ago, I limped noticeably. Casual acquaintances or people I saw infrequently came to identify me with that limp. Now they are happily surprised to see me walking normally. Their remarking this usually startles me. I never really thought of myself as being a cripple, except in a few dark hours. Whatever my physical condition is at any given time, that condition is normal. The fact that a person can get used to almost anything is a problem of our species. I was briefly accustomed to limping and now I'm reaccustomed to my wider range of capabilities, although a late-breaking career as a Rockette, or a newfound hobby in bungee jumping, are no longer among my options. What happened between the hobbling and the hikes in the woods? There was time. There was a genetic endowment of strength and health. There was abundant, wholesome nourishment. There was challenging, congenial work to do. There was an uncommon amount of freedom for me to set my own rhythm of work and rest, with no one I had to answer to. There was good medicine and surgical skill, homeopathic remedies to foster the reunion of the ends of the fracture, to prompt my body's own healing processes. Just living my blessedly privileged everyday life was healing, in the main. All that mercy was needed. I don't take having a full set of legs for granted.

The thing about a big injury—physical or psychic or ecological— is that organisms are not the same afterward. Hurt, loss, suffering, and recovery all work irrevocable changes. The question is whether or not one admits the fact of things being different now, and how one wears the difference. Tragedy is an overworked word. It's too often invoked to describe situations that are terrible indeed, but devoid of moral struggle and inexorability. I would not call that car accident of mine a tragedy. I do hope that it deepened me and may have precipitated a little wisdom. I might wish to be the person I was before the collision, the scalpels, the Demerol, the bouquets, the crutches, the pity, but she's long gone. And my right leg, although not a pretty sight, did do fifty percent of pedaling the bike the three miles over to the lake and back for much of the past two summers. In this case beauty is as beauty does. The asphalt ribbons down which I rode lay across a pastoralized countryside of considerable charm.

Still I know that the magnitude of injuries done to the land in the last century or so is so great that no matter what we do to restore it, it will never be the same—and by that I don't mean static, but equally complex—again. And, I think, the role that humans have played in ecosystems qualifies as tragic in the truest sense of the word. Perhaps the injured land's wisdom is that the human species has thus far been given a tragic role in the drama of evolution, like Macbeth. Whether it's inherent in us to continue this way is debatable. It remains possible for human beings and societies to change, especially in response to the parable of the living world which, after all, informed the majority of our experience as a species. Whatever the sign—the howl of a wolf on an island in Lake Superior, a dandelion pushing up through the asphalt at the edge of a playground, a bluebird taking up residence in a nesting box—we are evolved to note it, to make a reckoning of our fellow creatures, even as we all exist now, in conditions of extreme disturbance.

One might say that a great many human societies have treated the Earth narcissistically—being solely interested in what ecosystems and their constituent organisms could do to satisfy human needs, battening on the beauties and bounties of the Earth without ever really loving it for the self-generating marvel of its being. If one is the least bit willing to concede that there may be spirit and identity in nonhuman nature, the pain that it may feel for not being heard or seen or respected or cherished must by now be immense: the world is quite as wounded as we are, by the determined selfishness of people patterned by industrial civilization.

The land I look out from has suffered. It was dealt with like a first wife, the wife who after time grew tired and haggard and was less inclined to give at all. The virgin greenwood down the road beckoned richly and softly, and the husbandman departed, having been frustrated in his first attempt to secure a yield of a particular kind of comfort and joy. The author Wendell Berry says that the story of our time is one of divorce, and I fear that he is correct. We treat the Earth as we treat our affiliations. We demand outlandish things of it before we ever pause to listen, and there is no love in this.

In thinking about the Shack and the Arboretum, about the work in Auroville, and on the Mattole, and the North Branch prairies, one must be struck by what seems to be altruism, or an extraordinary imaginative empathy with nonhuman nature. To move to India and spend twenty years reclaiming a laterite plateau, to devote one's adulthood to the preservation, in context, of a native race of salmon, or to find a vocation in resurrecting within city limits a whole forgotten ecosystem with its membership of rare and endangered plants—these are, given the dominant (but crumbling) paradigm, peculiar, nonsensical behaviors. The economics are unusual, for one thing, and for another, the beneficiaries of this seeming altruism—Carnatic bird species reinhabiting their range, king salmon fry sheltered in a hatch box, the fringed prairie orchid germinating for the first time in a century in ancestral soils—do not repay their benefactors with ego-boosting thanks and praise, or even social justification.

However, the people I know who do this kind of work are satisfied by it in some basic and profound ways. It may be that they have developed what some thinkers are characterizing as an "ecological self," a sense of the human as being continuous with Nature, which, after all, is the simple truth. It's just that the political, economic, religious, and social forms that have come to predominate have required that we put the blinders on, and limit our concerns to self, state, and our own species. Ecological restoration is an act well-suited to rooting up this weedy, invasive paradigm and to reestablishing—with the significantly new amendment of conscious choice—ecosystems where ecological selves can make themselves at home and reproduce psychologically, if seldom physically.

It remains astonishing that people are willing to bend their backs working under punishing tropical sun or wade into icy rivers not for personal gain, but out of devotion. They are the bodhisattvas among us, and their numbers are growing. Someday, I hope, humanity will be a lot like them. This expansion or evolution of the human soul, its reintegration with the specifics of the planet, is an obvious and sensible next step, the lifesome move. It is quite a threshold to be poised upon.

Restoration is what lies before us, but the restoration must be of the whole system, and that whole ecosystem includes the human self, the personal heart. The transformative power of the great romance—be it with an admirable mate or a noble cause—remains marvelous. Surely the possibilities of it linger, surely as the seed bank of the forest will reproduce the woods, given the least advantage. If I could wish the total and complete restoration of the world, and the banishing of despair, I would wish for the immediate preservation of all species, all prairies, all forests, all swamps, all deserts; and for a return of crazy love, of go-for-broke passion between women and men, men and men, women and women, humans of all ages and places; between humans and soil and everything that arises therefrom. Let the love and commitment between beings be part of this great healing, and purify us of cynicism. If only we can dare to belong to one another, and to our land.

We have come to a moment in time when a wounded but willing humanity may look to healing itself in the land. In the individual, as well as in the ecological community, restoration begins with recapitulating the history in order to heal the land in the light of truth. Whatever sort of self we bring to the meeting with the wounded land, we may gain heart as we witness the Earth's response to care. The process is not solely in our hands; we have a part to play, but the power to restore belongs to Nature. We can only be abettors, not inventors in this endeavor; ours to defer to the greater art. Ecosystems are the greatest teachers. If we can approach a particular place humbly, attentively, openly, and with hopeful anticipation, if we earnestly try to discern what that place has wanted to be, what ecological community it has given rise to; if we try to be of real service to that place, we will see healing, and we will know love again. Conceivably the good feelings engendered within the Earth community, as restoration gains ground and beauty returns, could spiral wildly out of control and joy prevail.

We can be expansive enough to care about more than just our survival. As we approach the millennium, we may yet choose to realign our societies with our natural surroundings. Little by little. Yard by yard, stream by stream, acre by acre. The steps back in to the living world can be made, like Nature's own reclamation of an old field, in an organic succession, as the diverse and locally adapted community grows up, along with and around us. Neither our senses nor our ecological selves can be numbed or denied much longer. The sacred in Nature is finding a voice audible and intelligible to our moment. We can be freed from our tragic history to revive wildflowers we'll never pick, plant trees we'll never fell. We can grow whole in an emulation of Nature's disinterested generosity. It is our restoration to a timeless understanding, an immense knowledge. Now a new era can begin. A last chance and a first chance, for there was no golden age. We go out from the garden and into the wild.

Notes

Prologue

1. Paul R. Ehrlich and Anne H. Ehrlich, *Extinction: The Causes and Consequences of the Disappearance of Species* (New York: Random House, 1981), pp. xi–xiv.
2. Aldo Leopold, *A Sand County Almanac: With Essays on Conservation from Round River* (New York: Ballantine Books, 1974), p. 240.
3. In *A Green City Program for the San Francisco Bay Area and Beyond* (San Francisco: Planet Drum Books, 1989), Peter Berg and his colleagues Beryl Magilavy and Seth Zuckerman sketch a reinhabitory vision for San Francisco, one of America's great cities (see "Resources" section for access information on Planet Drum Books and other publications of the Planet Drum Foundation in which Berg's writings appear); Nancy Morita, "Wild in the City: The Ecology and Natural History of San Francisco" (San Anselmo, CA: Wild in the City, 1992). Morita titles her evocative presettlement map of San Francisco thusly. See "Resources" section for access information.
4. See "Resources" section for Turtle Island Office, which provides access information on the Turtle Island Bioregional Gatherings, formerly called the North American Bioregional Congresses.
5. Peter Berg and Raymond F. Dasmann, "Reinhabiting California," in *Reinhabiting a Separate Country: A Bioregional Anthology of Northern California*, ed. Pe-

ter Berg (San Francisco: Planet Drum Books, 1978), pp. 217–18. A salmagundi of Shastan reinhabitory culture.

Chapter 1. The Wild & The Tame

1. Stephanie Mills, *Whatever Happened to Ecology?* (San Francisco: Sierra Club Books, 1989).
2. Michael Soulé, "The Onslaught of Alien Species, and Other Challenges in the Coming Decades." *Conservation Biology*, 4(3) (September 1990):233. *Conservation Biology*, the journal of the Society for Conservation Biology, is an indispensable source for those wishing to stay abreast of developments in this "crisis discipline." See "Resources" section for access information.
3. David Orr, *Ecological Literacy: Education and the Transition to a Postmodern World* (Albany, NY: State University of New York Press, 1992). David Orr's thinking on what ecological literacy might be, and how to cultivate it in our society, is given full and persuasive expression in this upstart book.
4. Reed F. Noss, "The Wildlands Project Land Conservation Strategy." *Wild Earth* (Special Issue 1992):13. Noss's pivotal ideas for conservation biology are developed and elaborated in a book he has co-authored with Allen Y. Cooperrider entitled *Saving Nature's Legacy: Protecting and Restoring Biodiversity* (Washington, DC: Island Press, 1994).
5. Nancy Newhall, *This Is the American Earth* (San Francisco: Sierra Club Books, 1992), p. 62.
6. Thomas Berry, *The Dream of the Earth* (San Francisco: Sierra Club Books, 1988), p. 11. Thomas Berry's profound, portentous, erudite work provides a lucid philosophic underpinning for the human reinhabitation of place—from the watershed to the cosmos.
7. See "Resources" section for access to The Wildlands Project, which is the leading advocate of this vision.
8. Berry, *Dream of the Earth*, 215.
9. Stephan A. Hoeller, "Armageddon: Apocalyptic Musings for the End of the Century." *Gnosis*, 21 (Fall 1991):13.
10. Jeremiah Gorsline and Freeman House elaborate on this concept in an essay of the same name in *Home! A Bioregional Reader*, ed. Van Andruss, Christopher Plant, Judith Plant, and Eleanor Wright (Philadelphia: New Society Publishers, 1990). *Home!* contains many of the foundational texts of bioregionalism.

Chapter 2. Scraps of the Virgin Cloak

Background for this chapter came from *Atlas of Great Lakes Indian History* (Norman, OK: University of Oklahoma Press, 1986), edited by Helen Tanner; Betty Flanders Thomson's *The Shaping of America's Heartland: The Landscape of the Middle West* (Boston: Houghton Mifflin, 1977), a superb, sweeping physiography of the Middle West; and Susan L. Flader's symposium on the history of the Great

Lakes Forest, *The Great Lakes Forest: An Environmental and Social History* (Minneapolis, MN: University of Minnesota Press, 1983).

1. John A. Dorr and Donald F. Eschman, *Geology of Michigan* (Ann Arbor, MI: University of Michigan Press, 1970).

2. Lee Botts and Bruce Krushelnicki, *The Great Lakes: An Environmental Atlas and Resource Book* (Toronto: Environment Canada; and Chicago: U.S. Environmental Protection Agency, in cooperation with Brock University, Saint Catharines, Ontario, Canada, and Northwestern University, Evanston, Illinois, 1987), p. 4. See "Resources" section for information on obtaining (*gratis!*) a copy of this handy, comprehensive atlas which graphically details the environmental situation in the Great Lakes Basin.

3. Glenda Daniel and Jerry Sullivan, *The North Woods: A Sierra Club Naturalist's Guide* (San Francisco: Sierra Club Books, 1981), p. 134. This readable, richly informative guide is my nominee for the single most useful book on this region's natural history.

4. Robert E. Funk, "Post-Pleistocene Adaptations," in *Handbook of North American Indians*. Vol. 15, *Northeast*, ed. Bruce G. Trigger (Washington, DC: Smithsonian Institution, 1978), pp. 16–19. Other articles in this volume also gave helpful pedagogy on the Odawa, and on the prehistory of the Great Lakes region.

5. Charles E. Cleland, *Rites of Conquest: The History and Culture of Michigan's Native Americans* (Ann Arbor, MI: University of Michigan Press, 1992), pp.15–17.

6. Richard Asa Yarnell, *Aboriginal Relationships between Culture and Plant Life in the Upper Great Lakes Region* (Anthropological Papers 23, Museum of Anthropology. University of Michigan, Ann Arbor, Michigan, 1964), pp. 147–49.

7. James A. Clifton, George L. Cornell, and James M. McClurken, *People of the Three Fires: The Ottawa, Potawatomi and Ojibway of Michigan* (Grand Rapids, MI: Grand Rapids Inter-Tribal Council and The Michigan Indian Press, 1986), pp. 3–5.

8. Robert O. Petty, Essay and species notes in *Wild Plants in Flower*. Vol. 3, *Eastern Deciduous Forest* (Evanston, IL: Torkel Korling, 1977). An exquisite volume of prose and photography chronicling the wildflowers of this forest.

9. Nancy Nowak Cleland, *Vegetation as Recorded on Surveys of 1839*, Map 2154 (Michigan Department of Natural Resources in cooperation with Michigan State University, East Lansing, Michigan, March 1973).

10. Rutherford Platt, *The Great American Forest* (Englewood Cliffs, NJ: Prentice-Hall, 1966), pp. 57–59.

11. Dennis A. Albert and Leah D. Minc, "The Natural Ecology and Cultural History of the Colonial Point Red Oak Stands." A Report to the Michigan Department of Natural Resources and the University of Michigan Biological Station, Pellston, Michigan, March 1987.

12. Platt, *Great American Forest*, 136.

13. Bruce Catton, *Michigan: A Bicentennial History* (New York: W. W. Norton, 1976), p. 105.

14. See "Resources" section for Northwoods Wilderness Recovery, Inc., which provides information on the Michigamme Highlands campaign.

Chapter 3. Woods, Woods, and Nothing but Woods

1. Edmund M. Littell, *100 Years in Leelanau: A History of Leelanau County, Michigan* (Leland, MI: The Print Shop, 1965), p. 57.
2. Leelanau Township Historical Writers Group, *A History of Leelanau Township* (Leland, MI: Friends of the Leelanau Township Library, 1983), p. 124.
3. Aldo Leopold, *A Sand County Almanac: With Essays on Conservation from Round River* (New York: Ballantine Books, 1974), p. 118.
4. Bruce Catton, *Michigan: A Bicentennial History* (New York: W. W. Norton, 1976), p. 8.
5. William N. Sparhawk and Warren D. Brush, *The Economic Aspects of Forest Destruction in Northern Michigan*. Technical Bulletin 92 (U.S. Department of Agriculture, Washington, DC, 1929), pp. 9–11.
6. Rob Nurre's one-man show on the survey process, "The Surly Surveyor: A Look at the Pre-Settlement Landscape of the Old Northwest Territory through the Eyes of the Early Land Surveyors, 1785–1907" (Acorn Land Resources, Stevens Point, Wisconsin, 1990), presented at a restoration conference, was a lively introduction to this saga; see also Rutherford H. Platt, Jr., *Land Use Control: Geography, Law, and Public Policy* (Englewood Cliffs, NJ: Prentice-Hall, 1991), pp. 257–63.
7. Kasson Heritage Group, *Remembering Yesterday* (Kasson Township, Leelanau County, Michigan), pp. 7–9.
8. Robert Doherty, personal communication, June 17, 1994.
9. Ibid.
10. William James Gribb, "The Grand Traverse Bands' Land Base: A Cultural Historical Study of Land Transfer in Michigan" (Ph.D. thesis, Michigan State University, East Lansing, Michigan, 1982).
11. Murray Holliday, "The Leland Lake Superior Iron Works" (unpublished manuscript, Leland, Michigan).
12. Edward O. Wilson, "Is Humanity Suicidal?" (*New York Times Magazine*, May 30, 1993), p. 27. A preeminent conservation biologist, Wilson is also a writer of considerable grace. His book *The Diversity of Life* (Cambridge, MA: The Belknap Press of Harvard University Press, 1992) is a paen to, and primer on, the evolution of biodiversity, as well as being authoritative and appalling on the subject of the multifarious threats to same.

Chapter 4. Disturbed Ground

1. Aldo Leopold, "Wilderness," in *The River of the Mother of God and Other Essays by Aldo Leopold*, ed. Susan L. Flader and J. Baird Callicott (Madison, WI: University of Wisconsin Press, 1991), pp. 228–29.
2. Z. Obmiński, "Ecological Aspects"; T. Przybylski, "Morphology"; in *Scots Pine*

– *Pinus sylvestris L.* (Published for the U.S. Department of Agriculture and the National Science Foundation, Washington, D.C., by the Foreign Scientific Publications Department of the National Center for Scientific, Technical and Economic Information, Warsaw, Poland, 1975). Pryzbylski, Obmiński, and the 11th edition of the *Encyclopaedia Britannica* taught me every intellectual thing I know about Scotch, or Scots, pine.

3. From Emerson's *Fortune of the Republic*, as quoted in Bartlett's *Familiar Quotations*, 15th (125th anniversary) edition, ed. Emily Morison Beck (Boston: Little, Brown, 1980), p. 500.

4. Henry Allan Gleason, *The Plants of Michigan: Simple Keys for the Identification of the Native Seed Plants of the State* (Ann Arbor, MI: George Wahr, 1939).

5. Lauren Brown, *Grasses: An Identification Guide* (Boston: Houghton Mifflin, 1979).

6. Alfred W. Crosby, *Ecological Imperialism: The Biological Expansion of Europe, 900–1900* (Cambridge, England: Cambridge University Press, 1986), pp. 2–3. A fascinating history of the way in which European colonialism both deliberately and inadvertently produced ecological conditions favorable (at least in the short term) to itself.

7. Pjerre Dansereau, *Biogeography: An Ecological Perspective* (New York: Ronald Press, 1957), p. 160.

8. Rollin H. Baker, *Michigan Mammals* (East Lansing, MI: Michigan State University Press, 1983), pp. 137–46.

9. Baker, *Michigan Mammals*, 226–34.

10. Baker, *Michigan Mammals*, 390–99.

11. Baker, *Michigan Mammals*, 20–26.

12. Donald Worster, *Nature's Economy: A History of Ecological Ideas* (Cambridge, England: Cambridge University Press, 1977), pp. 73–74.

Chapter 5. The Leopolds' Shack

1. Curt Meine, *Aldo Leopold: His Life and Work* (Madison, WI: University of Wisconsin Press, 1988); Susan L. Flader, *Thinking Like a Mountain: Aldo Leopold and the Evolution of an Ecological Attitude toward Deer, Wolves, and Forests* (Columbia, MO: University of Missouri Press, 1974). Meine's biography is comprehensive and nicely written. Flader's is focused on the intellectual history of Leopold's ideas on deer, wolves, and forests, and is rigorous, insightful, and illuminating.

2. Aldo Leopold, *A Sand County Almanac: With Essays on Conservation from Round River* (New York: Ballantine Books, 1974), p. 190.

3. Leopold, *A Sand County Almanac*, 237–39.

4. Leopold, *A Sand County Almanac*, 197.

5. Alfred G. Etter, "A Day with Aldo Leopold," in *From the Land*, ed. Nancy P. Pittman (Washington, DC: Island Press, 1988), pp. 384–88.

6. Leopold, *A Sand County Almanac*, 103.

7. From *The River of the Mother of God and Other Essays by Aldo Leopold*, ed. Susan L. Flader and J. Baird Callicott (Madison, WI: University of Wisconsin Press, 1991), p. 191.

Chapter 6. Learning Restoration

1. Nancy Sachse's *A Thousand Ages: The University of Wisconsin Aboretum* (Regents of the University of Wisconsin, Madison, Wisconsin, 1974) is a useful history of the Arboretum, from the Ice Age to the mid-1960s. It includes lists of plant and animal species found there. See "Resources" section for access information on the Arboretum.
2. Curt Meine, *Aldo Leopold: His Life and Work* (Madison, WI: University of Wisconsin Press, 1988), p. 312.
3. From *The River of the Mother of God and Other Essays by Aldo Leopold*, ed. Susan L. Flader and J. Baird Callicott (Madison, WI: University of Wisconsin Press, 1991), pp. 210–11.
4. Robert E. Grese, "Historical Perspectives on Designing with Nature," in *Restoration '89: The New Management Challenge*, ed. H. Glenn Hughes and Thomas M. Bonnicksen (Proceedings of the First Annual Meeting of the Society for Ecological Restoration, Madison, Wisconsin, 1990), pp. 43–44.
5. Dave Egan, "Historic Initiatives in Ecological Restoration," *Restoration and Management Notes*, 8(2) (Winter 1990):83.
6. Virginia M. Kline, "How Well Can We Do? Henry Greene's Remarkable Prairie," *Restoration and Management Notes*, 10(1) (Summer 1992):36–37.
7. See "Resources" section for information on joining SER and receiving *Restoration and Management Notes*, a layperson-friendly journal which offers feature articles reporting on restoration projects around the world, reviews, announcements, and abstracts. *The* periodical in the field.

Chapter 7. Prairie University

1. Steve Packard's "Just a Few Oddball Species" engagingly recounts his savanna species sleuthing. It is found in *Helping Nature Heal: An Introduction to Environmental Restoration* (Berkeley, CA: Ten Speed Press/Whole Earth Catalog, 1991), edited by Richard Nilsen. *Helping Nature Heal* is a lively, wide-ranging, down-to-earth collection of articles on restoration projects and reviews of tools—conceptual and physical—useful to citizen-restorationists.
2. See "Resources" section for access information on Prairie University.
3. *Growing Native* is a newsletter that advises home gardeners in Alta California on landscaping with indigenous plants. See "Resources" section for access information.
4. Laurel Ross, "The Deer Problem," *North Branch Prairie Project Brush Piles* (Winter 1991–1992).

5. See "Resources" section for access information on the Volunteer Stewardship Network and its newsletters.

Chapter 8. Salmon Support

Phillip Johnson's article "Salmon Ranching: Can Fish Be Branded at Birth?" in *Oceans* (January 1982) intelligently appraised the drawbacks of hatcheries, thus was useful background for this chapter; also, Roderick Haig-Brown's classic nature saga, *Return to the River* (New York: Crown, 1946), was detailed and inspirational, a good read.

1. Janet Morrison, "Landforms, Waterflow, and People of the Mattole Watershed," in *Elements of Recovery: An Inventory of Upslope Sources of Sedimentation in the Mattole River Watershed with Rehabilitation Prescriptions and Additional Information for Erosion Control Prioritization* (prepared by the Mattole Restoration Council for the California Department of Fish and Game, Petrolia, California, December 1989), p. 12; see "Resources" section for access information on the Mattole Restoration Council.

2. The Mattole Watershed Salmon Support Group "keeps its feet wet," says Morrison. The group attends to the fish in the river, while the restoration council concerns itself more with the Mattole watershed's uplands. See "Resources" section for access information on the Mattole Watershed Salmon Support Group.

3. Quoted by Freeman House "working from A. L. Kroeber: *Indians of California*" in House's magisterial essay, "Totem Salmon," in *Home! A Bioregional Reader*, ed. Van Andruss, Christopher Plant, Judith Plant, and Eleanor Wright (Philadelphia: New Society Publishers, 1990), p. 69.

4. Erna Gunther, *A Further Analysis of the First Salmon Ceremony*, University of Washington Publications in Anthropology (Seattle, WA: University of Washington Press, June 1928), p. 140.

5. Tom Jay, "Salmon of the Heart," in *Working the Woods Working the Sea: Dalmo' Ma VI, An Anthology of Northwest Writings*, ed. Finn Wilcox and Jeremiah Gorsline (Port Townsend, WA: Empty Bowl, 1986), p. 102. "Salmon of the Heart" follows on House's "Totem Salmon," and is a remarkable essay, weaving together poems, myths, etymologies, and fish stories to limn the psyche of the salmon.

6. James Ludwig, "Wildlife Bioeffects and Toxic Chemicals in Great Lakes Fish: Implications for Human Health Effects" (presentation at Backyard Eco Conference '91, Lake Station, Michigan, 1991); for a responsible (but nonetheless alarming) general scientific survey of the condition of the Great Lakes ecosystem, *Great Lakes: Great Legacy?*, edited by Theodora E. Colborn, Alex Davison, Sharon N. Green, R. A. Hodge, C. Ian Jackson, and Richard A. Liroff (Washington, DC: Conservation Foundation; Ottawa, ONT: Institute for Research on Public Policy, 1990), is peerless, and makes reference to Ludwig's research as well as that of legions of other scientists.

7. Freeman House, personal communication, October 1993.

8. Freeman House, "To Learn the Things We Need to Know: Engaging the Particulars of the Planet's Recovery," in *Helping Nature Heal: An Introduction to Environmental Restoration*, ed. Richard Nilsen (Berkeley, CA: Ten Speed Press/ Whole Earth Catalog, 1991), p. 48.
9. "The Honeydew Slide Chronicles," in the *Mattole Restoration Newsletter*, 5 (Winter 1986), p. 3; see "Resources" section for access information.
10. Freeman House and David Simpson, *Mattole Restoration Newsletter*, 1 (1983), p. 4.
11. David Simpson, personal communication, November 1992.

Chapter 9. Greenwork in Utopia

Madhav Gadgil and Ramachandra Guha's *This Fissured Land: An Ecological History of India* (Oxford, England: Oxford University Press, 1992) not only provides a history of India's ecology, but also adduces a general theory of environmental history. It is a work of tremendous intellectual significance, and of great interest to any traveler in India.
1. Alan Lithman, "Revisiting Auroville: A Twenty-Year Project That's Still Growing," in *Helping Nature Heal: An Introduction to Environmental Restoration*, ed. Richard Nilsen (Berkeley, CA: Ten Speed Press/Whole Earth Catalog, 1991), p. 94.
2. Eric, "Report from Auroville," *The Coevolution Quarterly* (Spring 1974):72. The *Coevolution Quarterly* is now the *Whole Earth Review*, and continues reporting on ecological restoration as well as much else of interest (see "Resources" section for access information).
3. Quoted in "Introduction to Auroville" (pamphlet), published by Auroville International, U.S.A.; see "Resources" section for access information. The works of Sri Aurobindo—on Indian culture, philosophy, yoga, and nationalism—with his literary and poetic writings, run to sixty-six volumes; the works of the Mother comprise twelve volumes. All are published by the Sri Aurobindo Ashram Press in Pondicherry, India.
4. The three pamphlets were produced by the Auroville Greenwork Resource Center. See "Resources" section for access information.
5. K. Balasubramanyan, *Biotaxonomical Studies of Marakkanam R.F., Coromandel Coast*, published by the Department of Botany, Annamalai University, Annamalai Nagar, Tamil Nadu, India, in 1977.
6. Joss Brooks, personal communication, November 1993.
7. Ed Giordano, personal communication, December 1993.
8. Quoted in "Report of Awareness Workshops for a Sustainable Future," organized by Auroville's Center for Scientific Research (Auroville, India, September 1992); see "Resources" section for access information.

Chapter 10. Devotional Exercise

1. See "Resources" section for access information on Swatcha Ganga.
2. Malcolm Margolin's book *The Earth Manual: How to Work on Wild Land with-out Taming It* (Berkeley, CA: Heyday Books, 1985) is sensible, good, and a pleasure to read, with clear instructions on back-forty restoration techniques and a deeply rooted understanding of the reason to strive for ecological authenticity in the work.

John Berger's work as a writer, editor, and organizer has helped stimulate interest in restoration (see the Bibliography for two of his publications); see also "Resources" section for access to Restoring the Earth, which Berger directs.

Bibliography

Albert, Dennis A., and Leah D. Minc. "The Natural Ecology and Cultural History of the Colonial Point Red Oak Stands." Report to the Michigan Department of Natural Resources and the University of Michigan Biological Station, Pellston, Michigan, March 1987.

Andruss, Van, Christopher Plant, Judith Plant, and Eleanor Wright, eds. *Home! A Bioregional Reader*. Philadelphia: New Society Publishers, 1990.

Auroville Greenwork Resource Center. "Common Trees of Auroville." Auroville, India, March 1990.

Auroville Greenwork Resource Center. "The Nursery." Auroville, India, March 1990.

Auroville Greenwork Resource Center. "Trees of Auroville." Auroville, India, n.d.

Baker, Rollin H. *Michigan Mammals*. East Lansing, MI: Michigan State University Press, 1983.

Balasubramanyan, K. *Biotaxonomical Studies of Marakkanam R.F., Coromandel Coast*. Department of Botany, Annamalai University, Annamalainagar, Tamil Nadu, India, 1977.

Berg, Peter, and Raymond F. Dasmann. "Reinhabiting California," in *Reinhabiting a Separate Country: A Bioregional Anthology of Northern California*, ed. Peter Berg. San Francisco: Planet Drum Books, 1978.

Berg, Peter, Beryl Magilavy, and Seth Zuckerman. *A Green City Program for the San Francisco Bay Area and Beyond*. San Francisco: Planet Drum Books, 1989.

221

Berger, John J., ed. *Environmental Restoration: Science and Strategies for Restoring the Earth*. Washington, DC: Island Press, 1990.

Berger, John J. *Restoring the Earth: How Americans Are Working to Renew Our Damaged Environment*. New York: Anchor Press/Doubleday, 1987.

Berry, Thomas. *The Dream of the Earth*. San Francisco: Sierra Club Books, 1988.

Botts, Lee, and Bruce Krushelnicki. *The Great Lakes: An Environmental Atlas and Resource Book*. Toronto: Environment Canada; and Chicago: U.S. Environmental Protection Agency, in cooperation with Brock University, Saint Catharines, Ontario, Canada, and Northwestern University, Evanston, Illinois, 1987.

Braun, E. Lucy. *Deciduous Forests of Eastern North America*. New York: Hafner Publishing, 1964.

Brown, Lauren. *Grasses: An Identification Guide*. Boston: Houghton Mifflin, 1979.

Catton, Bruce. *Michigan: A Bicentennial History*. New York: W. W. Norton, 1976.

Center for Scientific Research. "Report of Awareness Workshops for a Sustainable Future." Auroville, India: Center for Scientific Research, September 1992.

Cleland, Charles E. *Rites of Conquest: The History and Culture of Michigan's Native Americans*. Ann Arbor, MI: University of Michigan Press, 1992.

Cleland, Nancy Nowak. *Vegetation as Recorded on Surveys of 1839*. Map 2154. Michigan Department of Natural Resources in cooperation with Michigan State University, East Lansing, Michigan, March 1973.

Clifton, James A., George L. Cornell, and James M. McClurken. *People of the Three Fires: The Ottawa, Potawatomi and Ojibway of Michigan*. Grand Rapids, MI: Grand Rapids Inter-Tribal Council and The Michigan Indian Press, 1986.

Colborn, Theodora E., Alex Davison, Sharon N. Green, R. A. Hodge, C. Ian Jackson, and Richard A. Liroff, eds. *Great Lakes: Great Legacy?* Washington, DC: Conservation Foundation; Ottawa, ONT: Institute for Research on Public Policy, 1990.

Crosby, Alfred W. *Ecological Imperialism: The Biological Expansion of Europe, 900–1900*. Cambridge, England: Cambridge University Press, 1986.

Curtis, John T. *The Vegetation of Wisconsin: An Ordination of Plant Communities*. Madison, WI: University of Wisconsin Press, 1959.

Daniel, Glenda, and Jerry Sullivan. *The North Woods: A Sierra Club Naturalist's Guide*. San Francisco: Sierra Club Books, 1981.

Dansereau, Pierre. *Biogeography: An Ecological Perspective*. New York: Ronald Press, 1957.

Dorr, John A., and Donald F. Eschman. *Geology of Michigan*. Ann Arbor, MI: University of Michigan Press, 1970.

Eck, Diana L. *Banaras: City of Light*. Princeton, NJ: Princeton University Press, 1982.

Egan, Dave. "Historic Initiatives in Ecological Restoration." *Restoration and Management Notes*, 8(2) (Winter 1990).

Ehrlich, Paul R., and Anne H. Ehrlich. *Extinction: The Causes and Consequences of the Disappearance of Species*. New York: Random House, 1981.

Eric, "Report from Auroville." *The Coevolution Quarterly* (Spring 1974).

Etter, Alfred G. "A Day with Aldo Leopold," in *From the Land*, ed. Nancy P. Pittman. Washington, DC: Island Press, 1988.

Flader, Susan L., ed. *The Great Lakes Forest: An Environmental and Social History.* Minneapolis, MN: University of Minnesota Press, 1983.

Flader, Susan L. *Thinking Like a Mountain: Aldo Leopold and the Evolution of an Ecological Attitude toward Deer, Wolves, and Forests.* Columbia, MO: University of Missouri Press, 1974.

Flader, Susan L., and J. Baird Callicott, eds. *The River of the Mother of God and Other Essays by Aldo Leopold.* Madison, WI: University of Wisconsin Press, 1991.

Foreign Scientific Publications Department of the National Center for Scientific, Technical, and Economic Information, Warsaw, Poland. *Scots Pine – Pinus sylvestris L.* Washington, DC: U.S. Department of Agriculture and the National Science Foundation, 1975.

Funk, Robert E. "Post-Pleistocene Adaptations," in *Handbook of North American Indians. Vol. 15, Northeast*, ed. Bruce G. Trigger. Washington, DC: Smithsonian Institution, 1978.

Gadgil, Madhav, and Ramachandra Guha. *This Fissured Land: An Ecological History of India.* Oxford, England: Oxford University Press, 1992.

Gleason, Henry Allan. *The Plants of Michigan: Simple Keys for the Identification of the Native Seed Plants of the State.* Ann Arbor, MI: George Wahr, 1939.

Gorsline, Jeremiah, and Freeman House. "Future Primitive," in *Home! A Bioregional Reader*, ed. Van Andruss, Christopher Plant, Judith Plant, and Eleanor Wright. Philadelphia: New Society Publishers, 1990.

Grese, Robert E. "Historical Perspectives on Designing with Nature," in *Restoration '89: The New Management Challenge*, ed. H. Glenn Hughes and Thomas M. Bonnicksen. Proceedings of the First Annual Meeting of the Society for Ecological Restoration, Madison, Wisconsin, 1990.

Gribb, William James. "The Grand Traverse Bands' Land Base: A Cultural Historical Study of Land Transfer in Michigan." Ph.D. thesis, Michigan State University, East Lansing, Michigan, 1982.

Gunther, Erna. *A Further Analysis of the First Salmon Ceremony.* University of Washington Publications in Anthropology. Seattle, WA: University of Washington Press, June 1928.

Haig-Brown, Roderick. *Return to the River.* New York: Crown, 1946.

Hoeller, Stephan A. "Armageddon: Apocalyptic Musings for the End of the Century." *Gnosis*, 21(Fall 1991).

Holliday, Murray. "The Leland Lake Superior Iron Works." Unpublished manuscript, Leland, Michigan, n.d.

"The Honeydew Slide Chronicles." *Mattole Restoration Newsletter*, 5 (Winter 1986).

House, Freeman. "To Learn the Things We Need to Know: Engaging the Particulars of the Planet's Recovery," in *Helping Nature Heal: An Introduction to Environmental Restoration*, ed. Richard Nilsen. Berkeley, CA: Ten Speed Press/Whole Earth Catalog, 1991.

House, Freeman. "Totem Salmon," in *Home! A Bioregional Reader*, ed. Van Andruss, Christopher Plant, Judith Plant, and Eleanor Wright. Philadelphia: New Society Publishers, 1990.

Hughes, H. Glenn, and Thomas M. Bonnicksen, eds. *Restoration '89: The New Management Challenge*. Proceedings of the First Annual Meeting of the Society for Ecological Restoration, Madison, Wisconsin, 1990.

Jay, Tom. "Salmon of the Heart," in *Working the Woods Working the Sea: Dalmo' Ma VI, An Anthology of Northwest Writings*, ed. Finn Wilcox and Jeremiah Gorsline. Port Townsend, WA: Empty Bowl, 1986.

Johnson, Phillip. "Salmon Ranching: Can Fish Be Branded at Birth?" *Oceans* (January 1982).

Kasson Heritage Group. *Remembering Yesterday*. Kasson Township, Leelanau County, Michigan, n.d.

Kline, Virginia M. "How Well Can We Do? Henry Greene's Remarkable Prairie." *Restoration and Management Notes*, 10(1) (Summer 1992).

Leelanau Township Historical Writers Group. *A History of Leelanau Township*. Leland, MI: Friends of the Leelanau Township Library, 1983.

Leopold, Aldo. *Game Management*. Madison, WI: University of Wisconsin Press, 1986.

Leopold, Aldo. *A Sand County Almanac: With Essays on Conservation from Round River*. New York: Ballantine Books, 1974.

Lithman, Alan. "Revisiting Auroville: A Twenty-Year Project That's Still Growing," in *Helping Nature Heal: An Introduction to Environmental Restoration*, ed. Richard Nilsen. Berkeley, CA: Ten Speed Press/Whole Earth Catalog, 1991.

Littell, Edmund M. *100 Years in Leelanau: A History of Leelanau County, Michigan*. Leland, MI: The Print Shop, 1965.

Ludwig, James. "Wildlife Bioeffects and Toxic Chemicals in Great Lakes Fish: Implications for Human Health Effects." Presentation at Backyard Eco Conference '91, Lake Station, Michigan, 1991.

Margolin, Malcolm. *The Earth Manual: How to Work on Wild Land without Taming It*. Berkeley, CA: Heyday Books, 1985.

Mattole Restoration Council. *Elements of Recovery: An Inventory of Upslope Sources of Sedimentation in the Mattole River Watershed with Rehabilitation Prescriptions and Additional Information for Erosion Control Prioritization*. Prepared for the California Department of Fish and Game, Petrolia, California, December 1989.

Meine, Curt. *Aldo Leopold: His Life and Work*. Madison, WI: University of Wisconsin Press, 1988.

Mills, Stephanie. *Whatever Happened to Ecology?* San Francisco: Sierra Club Books, 1989.

Morita, Nancy. "Wild in the City: The Ecology and Natural History of San Francisco" (map). San Anselmo, CA: Wild in the City, 1992.

Morrison, Janet. "Landforms, Waterflow, and People of the Mattole Watershed," in *Elements of Recovery: An Inventory of Upslope Sources of Sedimentation in the*

Mattole River Watershed with Rehabilitation Prescriptions and Additional Information for Erosion Control Prioritization. Prepared for the California Department of Fish and Game by the Mattole Restoration Council, Petrolia, California, December 1989.

Newhall, Nancy. *This Is the American Earth.* San Francisco: Sierra Club Books, 1992.

Nilsen, Richard, ed. *Helping Nature Heal: An Introduction to Environmental Restoration.* Berkeley, CA: Ten Speed Press/Whole Earth Catalog, 1991.

Noss, Reed F. "The Wildlands Project Land Conservation Strategy." *Wild Earth* (Special Issue 1992).

Noss, Reed F. and Allen Y. Cooperrider. *Saving Nature's Legacy: Protecting and Restoring Biodiversity.* Washington, DC: Island Press, 1994.

Nurre, Rob. "The Surly Surveyor: A Look at the Pre-Settlement Landscape of the Old Northwest Territory through the Eyes of the Early Land Surveyors, 1785–1907." Acorn Land Resources, Stevens Point, Wisconsin, 1990.

Obmiński, Z. "Ecological Aspects," in *Scots Pine – Pinus sylvestris L.* Foreign Scientific Publications Department of the National Center for Scientific, Technical, and Economic Information, Warsaw, Poland. Washington, DC: U.S. Department of Agriculture and the National Science Foundation, 1975.

Orr, David. *Ecological Literacy: Education and the Transition to a Postmodern World.* Albany, NY: State University of New York Press, 1992.

Packard, Steve. "Just a Few Oddball Species," in *Helping Nature Heal: An Introduction to Environmental Restoration*, ed. Richard Nilsen. Berkeley, CA: Ten Speed Press/Whole Earth Catalog, 1991.

Petty, Robert O. *Wild Plants in Flower. Vol. 3, Eastern Deciduous Forest.* (Essay and species notes). Evanston, IL: Torkel Korling, 1977.

Platt, Rutherford. *The Great American Forest.* Englewood Cliffs, NJ: Prentice-Hall, 1966.

Platt, Rutherford H., Jr. *Land Use Control: Geography, Law, and Public Policy.* Englewood Cliffs, NJ: Prentice-Hall, 1991.

Przybylski, T. "Morphology," in *Scots Pine – Pinus sylvestris L.* Foreign Scientific Publications Department of the National Center for Scientific, Technical, and Economic Information, Warsaw, Poland. Washington, DC: U.S. Department of Agriculture and the National Science Foundation, 1975.

Ross, Laurel. "The Deer Problem." *North Branch Prairie Project Brush Piles* (Winter 1991–1992).

Sachse, Nancy D. *A Thousand Ages: The University of Wisconsin Arboretum.* Regents of the University of Wisconsin, Madison, Wisconsin, 1974.

Scots Pine – Pinus sylvestris L. Published for the U.S. Department of Agriculture and the National Science Foundation, Washington, DC, by the Foreign Scientific Publications Department of the National Center for Scientific, Technical, and Economic Information, Warsaw, Poland, 1975.

Soulé, Michael. "The Onslaught of Alien Species, and Other Challenges in the Coming Decades." *Conservation Biology*, 4(3) (September 1990).

Sparhawk, William N., and Warren D. Brush. *The Economic Aspects of Forest Destruction in Northern Michigan*. Technical Bulletin 92. U.S. Department of Agriculture, Washington, DC, 1929.

Sullivan, Bill. "Auroville: A Case Study in Human Ecology," in Report of Awareness Workshops for a Sustainable Future, Center for Scientific Research, Auroville, Tamil Nadu, India, September 1992.

Tanner, Helen, ed. *Atlas of Great Lakes Indian History*. Norman, OK: University of Oklahoma Press, 1986.

Thomson, Betty Flanders. *The Shaping of America's Heartland: The Landscape of the Middle West*. Boston: Houghton Mifflin, 1977.

Trigger, Bruce G., ed. *Northeast*, Vol. 15 of *Handbook of North American Indians*, ed. William C. Sturtevant. Washington, DC: Smithsonian Institution, 1978.

Wilcox, Finn, and Jeremiah Gorsline, eds. *Working the Woods Working the Sea: Dalmo' Ma VI, An Anthology of Northwest Writings*. Port Townsend, WA: Empty Bowl, 1986.

Wilson, Edward O. *The Diversity of Life*. Cambridge, MA: The Belknap Press of Harvard University Press, 1992.

Wilson, Edward O. "Is Humanity Suicidal?," in *New York Times Magazine*, May 30, 1993.

Worster, Donald. *Nature's Economy: A History of Ecological Ideas*. Cambridge, England: Cambridge University Press, 1977.

Yarnell, Richard Asa. *Aboriginal Relationships between Culture and Plant Life in the Upper Great Lakes Region*. Anthropological Papers 23. Museum of Anthropology, University of Michigan, Ann Arbor, Michigan, 1964.

Resources

Auroville Greenwork Resource Center
Isai Ambalam
Auroville 605101
Tamil Nadu, INDIA

Auroville International U.S.A.
P.O. Box 162489
Sacramento, CA 95816-2489

Auroville Secretariat
Bharat Nivas
Auroville 605101
Tamil Nadu, INDIA

Center for Scientific Research
Auroshilpam
Auroville 605101
Tamil Nadu, INDIA

Conservation Biology
a quarterly journal of The Society for

Conservation Biology, annual
membership (includes subscription) $49
Blackwell Scientific Publications
238 Main Street
Cambridge, MA 02142

Friends of the Ganges
annual membership (includes newsletter
subscription and updates on Ganges
events) $35
Friends of the Ganges
3181 Mission Street #30
San Francisco, CA 94110

The Great Lakes
copies may be obtained from:
Great Lakes National Program Office
U.S. Environmental Protection Agency
230 South Dearborn Street
Chicago, IL 60604

Resources

Growing Native
a bimonthly newsletter of
The Growing Native Research Institute
Louise Lacey, Editor and Publisher
annual membership (includes
subscription) $30
Growing Native
P.O. Box 489
Berkeley, CA 94701

Mattole Restoration Newsletter
Mattole Restoration Council
P.O. Box 160
Petrolia, CA 95558
you may become a "friend of the
Mattole," and support their work
for a mere $10

Mattole Watershed
Salmon Support Group
P.O. Box 188
Petrolia, CA 95558

Northwoods Wilderness Recovery, Inc.
is working to preserve the Michigamme
Highlands and other wildlands in the
Upper Peninsula and environs; for
information and to support their work,
contact:
Northwoods Wilderness Recovery, Inc.
P.O. Box 107
Houghton, MI 49931-0107

Planet Drum Books
Planet Drum Foundation
annual membership (includes
subscription to the twice-yearly
magazine *Raise the Stakes*) $20
P.O. Box 31251
San Francisco, CA 94131

Prairie University
for information on Prairie University,

and on the Volunteer Stewardship
Network and its newsletters, or to get
involved, contact:
The Nature Conservancy
Illinois Field Office
79 West Monroe Street, Suite 900
Chicago, IL 60603

*Restoring Forestry: An International
Guide to Sustainable Forestry Practices*
Michael Pilarski, Editor
Kivaki Press
585 East 31st Street
Durango, CO 81301

Restoring the Earth
1713C Martin Luther King, Jr. Way
Berkeley, CA 94709

Society for Ecological Restoration
annual membership (includes
subscription to twice-yearly *Restoration
and Management Notes*) $32
Society for Ecological Restoration
University of Wisconsin-Madison
Arboretum
1207 Seminole Highway
Madison, WI 53711

Swatcha Ganga
Sankat Mochan Foundation
Tulsi Mandir
Tulsi Ghat 221005
Varanasi, INDIA

Turtle Island Office
Learning Alliance
494 Broadway
New York, NY 10012
    or
Cress Spring Farm
4035 Ryan Road
Blue Mounds, WI 53517

Resources

University of Wisconsin-Madison
Arboretum
annual membership (includes monthly
newsletter) $15
Friends of the Arboretum
University of Wisconsin Arboretum
1207 Seminole Highway
Madison, WI 53711

Whole Earth Review
a quarterly journal available for $20/yr.
from
Whole Earth Review
P.O. Box 38
Sausalito, CA 94966

Wild Earth
a quarterly journal reports on the
Wildlands Project (and much, much else
besides) $25/yr. from
The Cenozoic Society
P.O. Box 455
Richmond, VT 05477

Wild in the City
6 Cypress Road
San Anselmo, CA 94960

The Wildlands Project
1955 West Grant Road, Suite 148A
Tucson, AZ 85745

Index